Lecture Notes in Economics and Mathematical Systems

622

Founding Editors:

M. Beckmann
H.P. Künzi

Managing Editors:

Prof. Dr. G. Fandel
Fachbereich Wirtschaftswissenschaften
Fernuniversität Hagen
Feithstr. 140/AVZ II, 58084 Hagen, Germany

Prof. Dr. W. Trockel
Institut für Mathematische Wirtschaftsforschung (IMW)
Universität Bielefeld
Universitätsstr. 25, 33615 Bielefeld, Germany

Editorial Board:

A. Basile, H. Dawid, K. Inderfurth, W. Kürsten

Stefan Rostek

Option Pricing in Fractional Brownian Markets

 Springer

Dr. Stefan Rostek
University of Tübingen
Wirtschaftswissenschaftliches Seminar
Lehrstuhl für Betriebliche Finanzwirtschaft
Mohlstraße 36
72074 Tübingen
Germany
stefan.rostek@uni-tuebingen.de

ISSN 0075-8442
ISBN 978-3-642-00330-1   e-ISBN 978-3-642-00331-8
DOI 10.1007/978-3-642-00331-8
Springer Dordrecht Heidelberg London New York

Library of Congress Control Number: "PCN applied for"

Cover design: SPi Publishing Services

Printed on acid-free paper

springer is part & Springer Science+Business Media (www.springer.com)

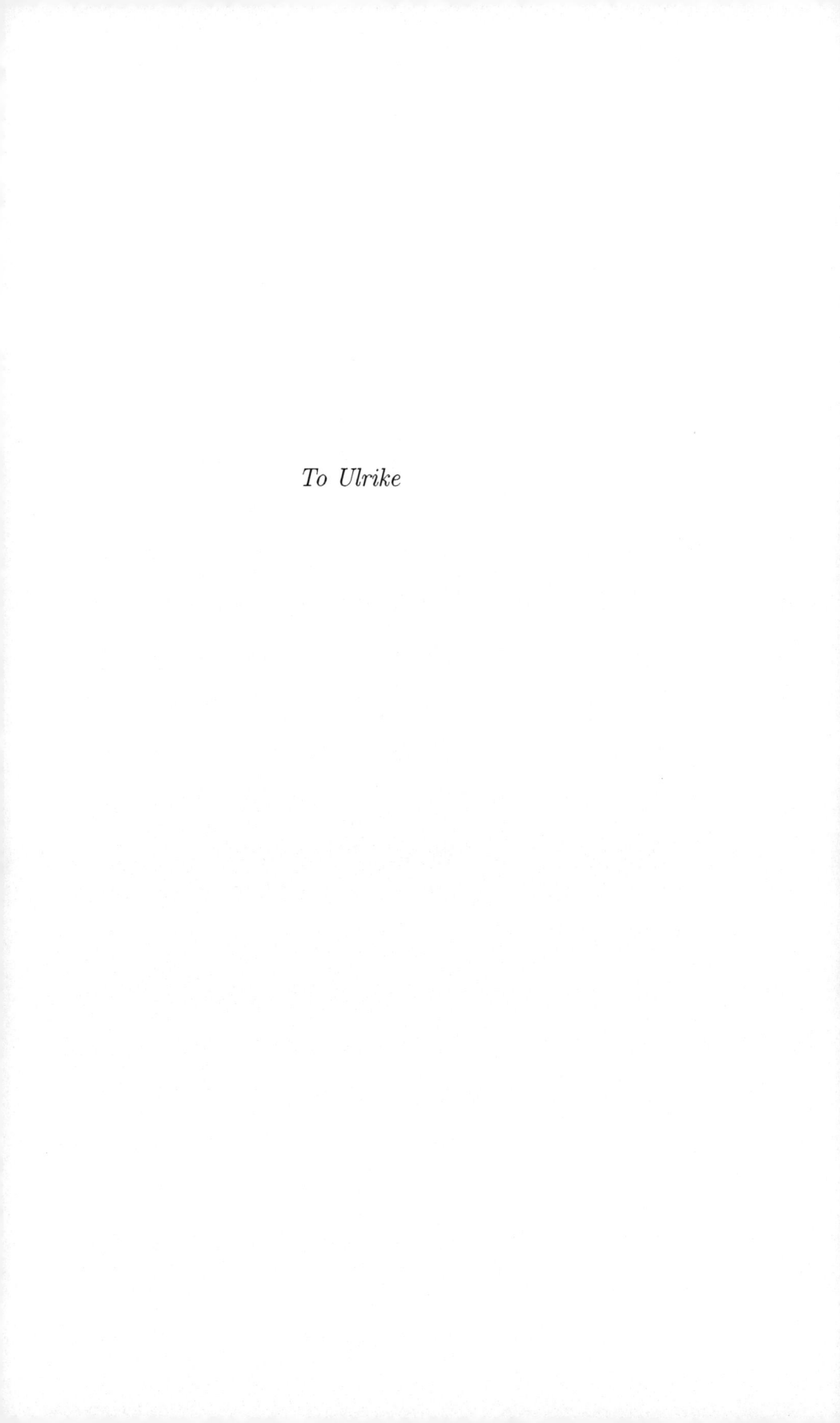

To Ulrike

Foreword

Mandelbrot and van Ness (1968) suggested fractional Brownian motion as a parsimonious model for the dynamics of financial price data, which allows for dependence between returns over time. Starting with Rogers (1997) there is an ongoing dispute on the proper usage of fractional Brownian motion in option pricing theory. Problems arise because fractional Brownian motion is not a semimartingale and therefore "no arbitrage pricing" cannot be applied. While this is consensus, the consequences are not as clear. The orthodox interpretation is simply that fractional Brownian motion is an inadequate candidate for a price process. However, as shown by Cheridito (2003) any theoretical arbitrage opportunities disappear by assuming that market participants cannot react instantaneously.

This is the point of departure of Rostek's dissertation. He contributes to this research in several respects: (i) He delivers a thorough introduction to fractional integration calculus and uses the binomial approximation of fractional Brownian motion to give the reader a first idea of this special market setting. (ii) Similar to the classical work of Sethi and Lehoczky (1981) he compares Wick-Itô and Stratonovich integration for the unrestricted fractional Brownian case, obtaining deterministic option prices. This disproves in an elegant way several option pricing formulæ under fractional Brownian motion in the literature. (iii) If market prices move only slightly faster than any market participant can react, we are left with an incomplete market setting. Again, but now by a different reason, "no arbitrage pricing" cannot be applied. Based on Rostek and Schöbel (2006), he shows carefully and in great detail for the continuous as well as for the binomial setting that a risk preference based approach may be the solution to the option valuation puzzle under fractional Brownian motion.

I recommend this research monograph to everybody who is curious enough to learn more about the fragile character of our prevailing valuation theory.

Tübingen, December 2008 *Rainer Schöbel*

Acknowledgements

This book is the outcome of my three years lasting research work at the Department of Corporate Finance at the Eberhard Karls University of Tübingen. During this time I had the great fortune to be supported by a number of persons my heartfelt thanks go to. Moreover, I would like to single out the most important of these.

First and foremost, my thankfulness and appreciation are directed to my academic supervisor and teacher Prof. Dr.-Ing. Rainer Schöbel. His advice and guidance and particularly his scientific inquisitiveness accompanied by a perpetual positive mindset, heavily encouraged my work and formed the core of an ideal environment for my academic research. As a part of this stimulating environment, I would also like to thank Prof. Dr. Joachim Grammig for interesting discussions and advice, and not least for being the second referee of this thesis. Furthermore, my thanks go to my colleagues of the Corporate Finance Department Svenja Hager, Markus Bouziane, Robert Frontczak, Björn Lutz and Detlef Repplinger as well as Vera Klöckner for the friendly atmosphere and useful hints they provided.

I gratefully acknowledge the financial support of the Deutsche Forschungsgemeinschaft who funded my research as a member of the Research Training Group "Unternehmensentwicklung, Marktprozesse und Regulierung in dynamischen Entscheidungsmodellen".

I would like to express my deepest gratitude to my parents Roswitha and Franz Rostek. They were backing me all the way with their unrestricted faith in me and their enduring encouragement. Above all, I want to thank Ulrike Rostek, my beloved wife. Your patience, your understanding and your unconditional love are a godsend. Not knowing how to pay this off, I have to trust in Paul McCartney's 'fundamental theorem': "And in the end, the love you take is equal to the love you make." Thank you, you make everything so easy.

Schwieberdingen, December 2008   *Stefan Rostek*

Contents

Acronyms

σ	volatility parameter of the stock
μ	drift parameter of the stock
E	expectation operator
Var	variance operator
Cov	covariance operator
H	Hurst parameter
Γ	Gamma function
$\beta_{x,y}(z)$	incomplete Beta function
Ω	state space of random events
ω	random event or path
B_t^H	process of fractional Brownian motion at time t
t	current time
T	maturity time
$\mathbb{R}$	set of real numbers
B_t^H	process of Brownian motion at time t
τ	time to maturity
$\diamond$	Wick multiplier (diamond symbol)
$S(F)$	S-Transform of a function F
C_t	value of a European call option at time t
S_t	value of the basic risky asset at time t
$B_t^{H(n)}$	discrete n-step approximation of B_t^H
ξ	binomial random variable with zero mean and unit variance
n	number of discretization steps per unit of time
$\hat{B}_{T,t}^H$	conditional expectation of B_T^H at time t
$\hat{\sigma}_{T,t}^2$	conditional variance of B_T^H at time t
A_t	value of a deterministic bond
r	interest rate
K	strike price of a European option
S_0	initial price of the underlying of a European option
W_t^H	fractional White noise
$N(x)$	value of the standard normal distribution function

(I)	indicates Itô meaning of the following differential equation
(S)	indicates Stratonovich meaning of the following differential equation
(W)	indicates Wick-Itô meaning of the following differential equation
R_t	value of a dynamic portfolio at time t
P	probability measure on Ω
ρ_H	narrowing factor of the conditional distribution of fractional Brownian motion
fBm	fractional Brownian motion
$\hat{S}_t$	conditional stock price process
$\hat{B}_t$	conditional process of Brownian motion
$\bar{\mu}$	equilibrium drift rate
$\hat{B}_t^H$	conditional process of fractional Brownian motion
$\mathcal{F}_t$	information set available at time t
$I_{[\cdot,\cdot]}$	indicator function for a certain interval
η	partial derivative of the fractional call price with respect to the Hurst parameter H
ψ_0	digamma function

Chapter 1
Introduction

The vast majority of approaches towards option pricing deals with Brownian motion as a source of randomness. The seminal articles by Black and Scholes (1973) as well as by Merton (1973) crowned this evolution but did not conclude it by any means. Right up to today, the favorable properties and the well-developed stochastic calculus of classical Brownian motion attract both scientists and practitioners.

However, there was early evidence about some incompatibilities with regard to real market data. Concerning the stochastic process of Brownian motion, the main critique drawn from empiricism is at least two-fold:
On the one hand, real market distributions were shown to be not Gaussian (see e.g. Fama (1965)). The debate of recent years has put a great deal of effort on correcting this problem. Particularly the theory of Lévy processes allows it to incorporate a wide range of distributions into financial models. However, despite the large set of Lévy type stochastic processes, closed-form solutions are still limited to specific cases of non-Gaussian distributions. For more details about Lévy processes we refer the interested reader to the monograph of Cont and Tankov (2004) who provide a distinguished starting point to the topic.
On the other hand, the processes of observable market values seem to exhibit serial correlation (see e.g. Lo and MacKinley (1988)). Much less endeavor has been made to get a grip on this problem by factoring in aspects of persistence. However, at least there is one stochastic process that has often been proposed for mapping this kind of behavior: the very candidate is called fractional Brownian motion.

There are several reasons why we concern ourselves with this stochastic process. Fractional Brownian motion was originally introduced by Mandelbrot and van Ness (1968). It is a Gaussian stochastic process that is able to easily capture long-range dependencies or persistence. Being furthermore self-similar, its usage in financial models suggests itself. For reasons of parsimony,

S. Rostek, *Option Pricing in Fractional Brownian Markets.*
Lecture Notes in Economics and Mathematical Systems.
© Springer-Verlag Berlin Heidelberg 2009

we appreciate that fractional Brownian motion possesses only one additional parameter, the so-called Hurst parameter, which lies between one and zero. Over the range of parameter values, the process shows different shapes of inter-temporal correlation. Particular interest arises from the fact that the case of serial independence is included. Therefore, fractional Brownian motion is an extension of classical Brownian motion. Comparing the respective results will both feed intuition and allow for a checking of plausibility.

The fundamental question of this thesis is whether and to what extent one can draw parallels between the fractional and the classical Brownian motion framework. More precisely: As fractional Brownian motion is an extension of Brownian motion, is it also possible to extend the respective theory of option pricing? Are the well-developed techniques of stochastic calculus transferable to fractional Brownian motion? Will we be faced with conceptual problems? Can we obtain closed-form solutions?

We will tackle all these problems step-by-step. Several times, we will switch over from discrete to continuous time considerations and vice versa. The reason for this is the following: Certainly, one could strictly separate the respective discussions and treat the cases one by one. However, so doing and starting with the continuous time case, we would miss the opportunity to motivate the results by those of the more descriptive discrete time setting. Turning the tables, if we discussed the discrete time framework first, we could not check the approximation results by comparing them with their limit case. By contrast, the alternating argumentation provides the best possible mutual benefit of the two frameworks, and additionally enhances the readability of the thesis.

In our preliminary Chap. 2, we will recall and present the most important insights concerning fractional Brownian motion and the corresponding integration calculus. We will become acquainted with the typical characteristics of the process. Concerning integration theory, we will get to know different concepts. In particular, it will be the so-called Wick-based integration calculus that will provide us with fractional analogues to the fundamental results of the well-known Itô calculus.

To get a first idea about the fractional Brownian market setting and the appendant characteristics, we will deal in Chap. 3 with a binomial approximation of fractional Brownian motion. For reasons of illustration, we will depict fractional binomial trees. These trees will not only enhance understanding of distributional aspects of fractional Brownian motion, they will also indicate the main problem of fractional Brownian markets: In an unrestricted market setting, arbitrage opportunities can occur.

In Chap. 4 we will readdress ourselves with the continuous time case. The problem of arbitrage will be thoroughly discussed. After presenting the scientific debate of the history, we will clarify that the problem can be solved by restricting the set of feasible trading strategies. Motivated by the result from the discrete time framework, we will provide an elegant proof as to why a fractional Brownian market setting needs to be restricted. To this end, we will harness the reasoning of Sethi and Lehoczky (1981) and translate it into the fractional context. The result will be surprising at first glance but it will reveal perfectly the incompatibility of fractional Brownian motion and dynamical hedging. Consequently, we will renounce continuous tradability which is sufficient to ensure absence of arbitrage. As a proximate way out, we will suggest the transition to a risk preference based pricing approach.

Chapter 5 will form the core of this thesis and represents a further development of a preceding joint work by Rostek and Schöbel (2006). Assuming risk-neutral investors, we will price options in the continuous time fractional Brownian market. We will focus on a two-time valuation by postulating that the equilibrium condition we will introduce holds with respect to current time t and maturity T. We will apply some useful results concerning conditional expectation of fractional Brownian motion. Furthermore, we will state and use a conditional version of the fractional Itô theorem. Provided with these technical tools, we will be able to exploit the fundamental equilibrium condition. In the sense of a total equilibrium, the equilibrium condition will endogenously determine the drift of the underlying stock process. We will derive a closed-form solution for the price of a European option written on a stock that follows a fractional Brownian motion with arbitrary Hurst parameter H. Concerning the influence of the Hurst parameter H on the option price, we will elaborate different effects which we will call narrowing effect and maturity effect, respectively. Subsequently, we will consider the relation between option price and time to maturity which will lead us to the term structure of implied volatility. The latter will be a manifest result that clarifies the improvement our model yields.

By means of our derived results, we will be able to check how far appropriate results can also be drawn from our binomial approximation. In Chap. 6, we will therefore present the pricing approach from a discrete time vantage point. Like in the continuous time setting, we will first concentrate on a two-time valuation introducing a single equilibrium condition. We will address ourselves both to a relative and to an absolute equilibrium approach. Motivated by the ease and the traceability of the discrete time calculus, we will also consider multi-time equilibrium approaches. We will consider two different possibilities of stating the system of multi-time equilibrium conditions which will lead to totally different results. We will show that these results are in line with our understanding with respect to fractional Brownian motion.

We will finalize our dialectical consideration between discrete time and continuous time framework by making one further transition. In Sect. 6.4, we will use the deeper insight provided by the discrete multi-time results. In particular, we will ask ourselves what will happen if continuous time analogues of these multi-time equilibria are considered.

The concluding chapter summarizes the stated results and grants an outlook towards possible topics of further research.

Chapter 2
Fractional Integration Calculus

In order to model randomness in any stochastic model, one may do so by asserting a distribution of the random component. The somewhat more sophisticated approach—especially when modeling dynamical issues—is defining a suitable stochastic process. The overwhelming majority of treatable models based on stochastic processes deals with classical Brownian motion as the source of randomness. This is mainly due to the two main properties of this process, which are its Gaussian character, on the one hand, and its lack of serial correlation, on the other hand. However, though being easy to manage, these features often do not map things as they truly are. Real time series often fluctuate in a non-Gaussian fashion and/or are by all means serially correlated. A great deal of research effort has been invested to get a grip on the first problem; from the onset by introducing random jumps. Currently, researchers suggest so-called alpha-stable processes which are a special group of Levy processes. With the classical Brownian motion, these processes share the property of self-similarity.

However, in the literature of financial mathematics, few extensions have been proposed to overcome the assumption of independent increments for the stochastic processes. The most popular model was introduced by Mandelbrot and van Ness (1968). They hold true the Gaussian character of the process but allow for dependence over the line of time. Figure 2.1 by Cont and Tankov (2004) depicts the relations between important sets of stochastic processes. We see that while the intersection of all three sets is classical Brownian motion, fractional Brownian motion is still Gaussian and self-similar but no longer has independent increments.

From classical Brownian theory we learned that the transition from a deterministic framework to a stochastic one made it necessary to adjust the pertinent theory of integration. First of all, the definition of convergence had to be reconditioned in a mean square sense. Furthermore, the concept of a new integration calculus had to allow for the occurrence of infinite variation concerning the integrator. The celebrated solution to these problems was the

S. Rostek, *Option Pricing in Fractional Brownian Markets.*
Lecture Notes in Economics and Mathematical Systems.
© Springer-Verlag Berlin Heidelberg 2009

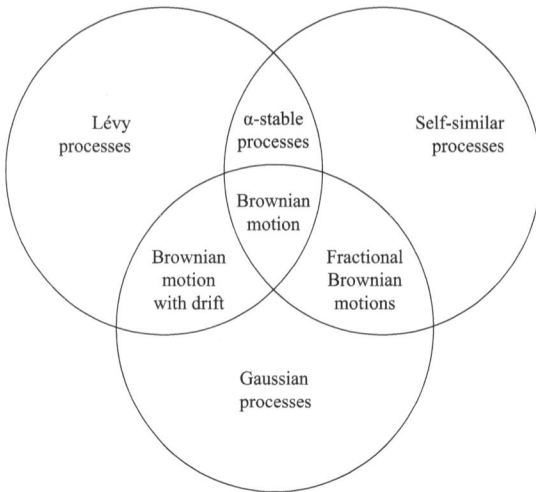

Fig. 2.1 Relations between different groups of stochastic processes (according to Cont/ Tankov (2006))

approach to stochastic integration for semimartingales by Itô (1951), consequently named Itô calculus.

It can be shown (see Rogers (1997), p. 3–4) that—except the case $H = \frac{1}{2}$— fractional Brownian motion is not a semimartingale. This rules out the application of conventional integration theory. In other words, making the transition to the stochastic process of fractional Brownian motion, the Itô integration theory itself becomes obsolete.

Several suggestions have been made in the past to overcome these problems and to extend the integration concept of Itô to a more general concept. This chapter presents the most important ones. We start with the investigation of the so-called Wick-based approach due to Duncan et al. (2000). It can be regarded as a milestone towards an integration theory with respect to fractional Brownian motion as it demonstrates existing parallels to classical theory and facilitates the development of a fractional Brownian market setup. Comparing the Wick-based approach to the alternative concept of a fractional integral of Stratonovich type, these advantages will become clear. We will briefly recall the main results of the Wick calculus, including fractional versions of well-known theorems. These findings culminate in a fractional Itô theorem which was provided by Duncan et al. (2000).

Yet, the Wick-based approach in its original version by Duncan et al. (2000) is limited to the persistent cases with Hurst parameters larger than one half. The transition to the antipersistent case can only successfully be made by means of another still more general integration concept, the S-transform approach by Bender (for an overview, see Bender (2003a)). We sketch the basic idea of this approach. It can be viewed as seminal with respect to a clean mathematical foundation of fractional integration theory.

The outline of this chapter is as follows. In the first section, we define fractional Brownian motion and highlight important properties. We then investigate the role of the so-called Hurst parameter and see how persistence or serial correlation comes into play. The remaining sections of the chapter will demonstrate this in a technical way. We introduce approaches to a stochastic calculus for fractional Brownian motion and present important parallels to classical Brownian theory as a fractional Girsanov theorem or a fractional Itô theorem.

2.1 The Stochastic Process of Fractional Brownian Motion

We use the definition of fractional Brownian motion via its original presentation as a moving average of Brownian increments. We introduce the following defining notation for this purpose:

$$(x)_+^y = \begin{cases} x^y & \text{if } x \geq 0, \\ 0 & \text{if } x < 0. \end{cases}$$

For $0 < H < 1$, fractional Brownian motion $\{B_t^H, t \in \mathbb{R}\}$ is the stochastic process defined by

$$B_0^H(\omega) = 0 \quad \forall \omega \in \Omega, \tag{2.1}$$

$$B_t^H(\omega) = c_H \left[\int_{\mathbb{R}} \left((t-s)_+^{H-\frac{1}{2}} - (-s)_+^{H-\frac{1}{2}} \right) dB_s(\omega) \right], \tag{2.2}$$

where $\{B_s, s \in \mathbb{R}\}$ is a two-sided Brownian motion, H is the so-called Hurst parameter and

$$c_H = \sqrt{\frac{2H\Gamma(\frac{3}{2} - H)}{\Gamma(\frac{1}{2} + H)\Gamma(2 - 2H)}}$$

is a normalizing constant. Note that for $t > 0$, B_t^H can be rewritten by

$$B_t^H = c_H \left[\int_{-\infty}^0 \left((t-s)^{H-\frac{1}{2}} - (-s)^{H-\frac{1}{2}} \right) dB_s + \int_0^t (t-s)^{H-\frac{1}{2}} dB_s \right].$$

Choosing the special parameter value $H = \frac{1}{2}$, we obtain

$$B_t^{\frac{1}{2}} = c_{\frac{1}{2}} \left[\int_{-\infty}^{0} \left((t-s)^{\frac{1}{2}-\frac{1}{2}} - (-s)^{\frac{1}{2}-\frac{1}{2}} \right) dB_s + \int_{0}^{t} (t-s)^{\frac{1}{2}-\frac{1}{2}} dB_s \right]$$

$$= \int_{0}^{t} dB_s = B_t,$$

where $\quad c_{\frac{1}{2}} = \sqrt{\dfrac{2 \cdot \frac{1}{2} \Gamma(\frac{3}{2} - \frac{1}{2})}{\Gamma(\frac{1}{2} + \frac{1}{2}) \Gamma(2 - 2 \cdot \frac{1}{2})}} = 1.$

Obviously, $B_t^{\frac{1}{2}}$ coincides with classical Brownian motion. On the other hand, in the next section, the cases $0 < H < \frac{1}{2}$ and $\frac{1}{2} < H < 1$ will be identified with the occurrence of antipersistence and persistence, respectively. Consequently, fractional Brownian motion can be divided into three families exhibiting—as we will see—quite different properties.

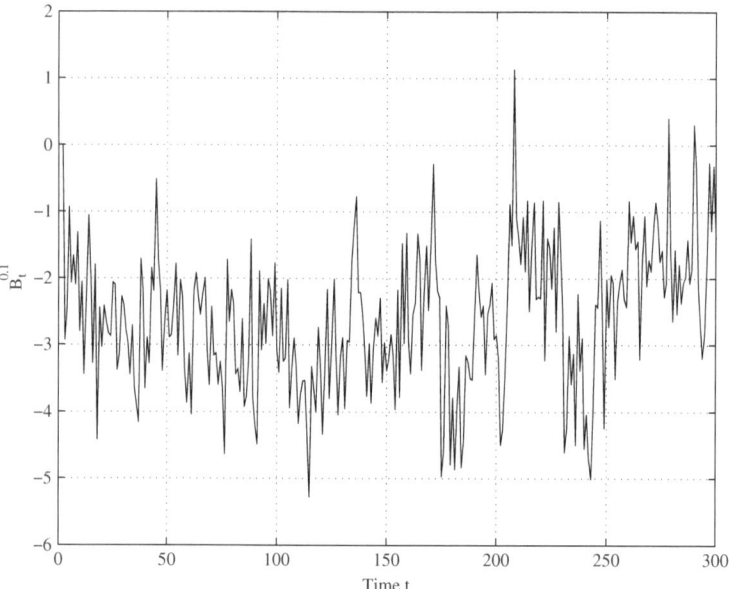

Fig. 2.2 Path of the fractional Brownian motion for $H = 0.1$

The Figs. 2.2–2.4 depict realizations of fractional Brownian motion. At first glance, we notice that the higher the Hurst parameter, the rougher the corresponding path. Looking on the scale of the axis, we also recognize that the smoother paths deviate considerably more from the zero mean. This is emphasized by Fig. 2.5 where the different processes are plotted in the same coordinate system. The next section will explain these phenomena in detail.

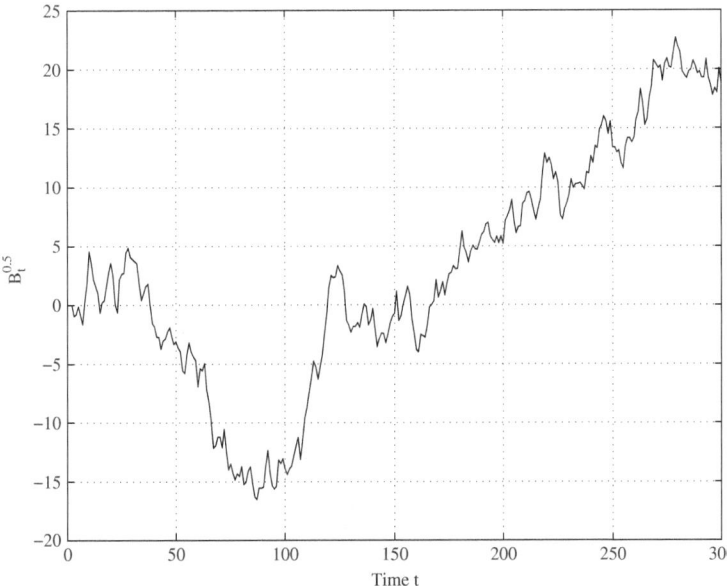

Fig. 2.3 Path of the fractional Brownian motion for $H = 0.5$

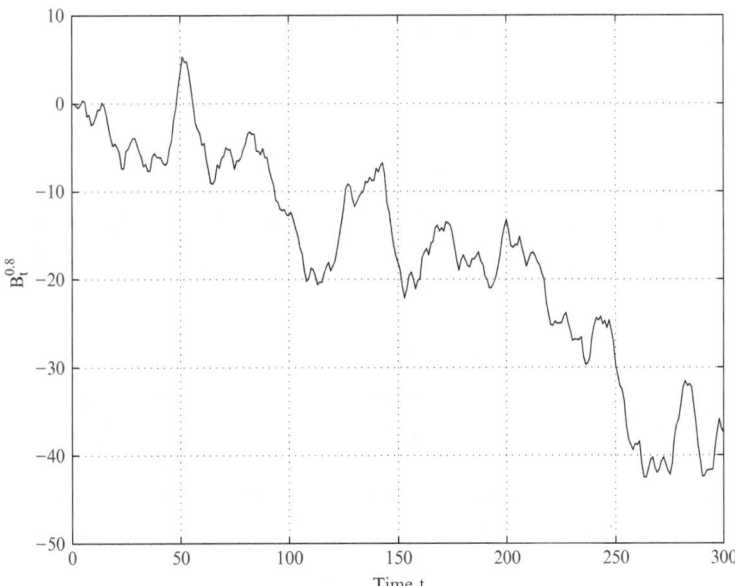

Fig. 2.4 Path of the fractional Brownian motion for $H = 0.8$

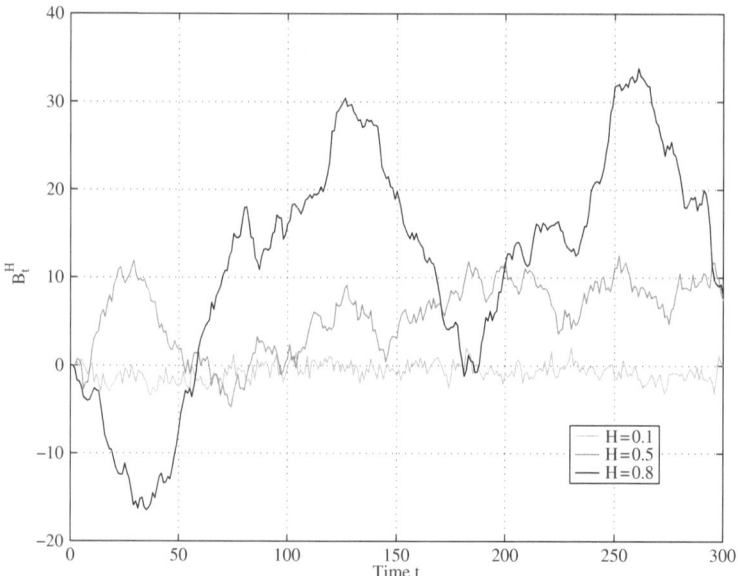

Fig. 2.5 Paths of the fractional Brownian motion for different Hurst parameters

In most of the recent articles concerning fractional Brownian motion, the stochastic process is not defined via its integral representation but—like classical Brownian motion—by its covariance properties. It can be shown, that B_t^H can also be represented as the unique, almost surely, continuous Gaussian process satisfying the following conditions:

$$E(B_t^H) = 0 \quad \forall\, t \in \mathbb{R},$$
$$E(B_t^H B_s^H) = \frac{1}{2}\left[|t|^{2H} + |s|^{2H} - |t-s|^{2H}\right], \ \forall\, t,s \in \mathbb{R}.$$

To verify that fractional Brownian motion indeed satisfies the covariance property above, we use a result concerning the expected value of products of fractional integrals with deterministic integrands. Note that when restricting ourselves to deterministic integrands only, no new integration theory has to be developed. We just postulate that the integrand can be approximated by a sequence of so-called simple functions $\sum_{i=1}^{n} \alpha_i I_{[t_i, t_{i+1}]}$ being piecewise constant. For a sequence of partitions π_n with $|\pi_n| \to 0$ of the interval $[0, t]$, the fractional integral with deterministic integrand can then be interpreted in the following sense:

$$\int_0^t f(s)\, dB_s^H := \lim_{n \to \infty} \sum_{\pi_n} \alpha_i (B_{t_{i+1}}^H - B_{t_i}^H).$$

Based on this definition, Gripenberg and Norros (1996) derive the following result:

$$E\left(\int_{\mathbb{R}} f(u) dB_u^H \int_{\mathbb{R}} g(v) dB_v^H\right) = H(2H-1) \int \int_{\mathbb{R}^2} f(u)g(v)|u-v|^{2H-2} du dv.$$

(2.3)

For the proof of this equation, see Gripenberg and Norros (1996).

With $f = I_{[0,t]}$ and $g = I_{[0,s]}$, and if we assume for example $t > s$, we get

$$\begin{aligned}
E\left(B_t^H B_s^H\right) &= E\left(\int_{\mathbb{R}} I_{[0,t]}(u) dB_u^H \int_{\mathbb{R}} I_{[0,s]}(v) dB_v^H\right) \\
&= H(2H-1) \int_0^s \int_0^t |u-v|^{2H-2} du\, dv \\
&= H(2H-1) \int_0^s \left(\int_0^v (v-u)^{2H-2} du + \int_v^t (u-v)^{2H-2} du\right) dv \\
&= H \int_0^s \left([-(v-u)^{2H-1}]_0^v + [(u-v)^{2H-1}]_v^t\right) ds \\
&= H \int_0^s \left(v^{2H-1} + (t-v)^{2H-1}\right) ds \\
&= \frac{1}{2} [v^{2H} - (t-v)^{2H}]_0^s \\
&= \frac{1}{2} [t^{2H} + s^{2H} - (t-s)^{2H}],
\end{aligned}$$

which finishes the proof of the statement.

Again, in the limit case $H = \frac{1}{2}$, the moment properties of classical Brownian motion can be obtained, as we receive for $t > s > 0$

$$E\left(B_t^{\frac{1}{2}} B_s^{\frac{1}{2}}\right) = \frac{1}{2}\left[|t|^1 + |s|^1 - |t-s|^1\right] = \frac{1}{2}[t + s - (t-s)] = s = \min(s,t).$$

As a further easy result, we get the variance of fractional Brownian motion

$$E((B_t^H)^2) = \frac{1}{2}\left[|t|^{2H} + |t|^{2H} - |t-t|^{2H}\right] = t^{2H} \quad \forall\, t,s \in \mathbb{R}.$$

From the covariance property, it follows that fractional Brownian motion itself is not a stationary process. But like classical Brownian motion, fractional Brownian motion does not reveal all its interesting properties until looking at its increments. We therefore look at the fractional Brownian increment

$$\Delta B_{t,s}^H = B_t^H - B_s^H,$$

which has the following moment properties:

$$E(\Delta B_{t,s}^H) = E(B_t^H - B_s^H) = E(B_t^H) - E(B_s^H) = 0, \ \forall \ t, s \in \mathbb{R},$$

$$E\left((\Delta B_{t,s}^H)^2\right) = E\left((B_t^H - B_s^H)(B_t^H - B_s^H)\right)$$
$$= E\left((B_t^H)^2\right) + E\left((B_s^H)^2\right) - 2E(B_t^H B_s^H)$$
$$= t^{2H} + s^{2H} - 2 \cdot \frac{1}{2}\left[|t|^{2H} + |s|^{2H} - |t-s|^{2H}\right]$$
$$= |t-s|^{2H} \quad \forall \ t, s \in \mathbb{R}.$$

Evidently, both the first and the second moment do not depend on the current point in time but only on the length of the increment: that is, fractional Brownian motion has stationary increments. However, in general, increments of fractional Brownian motion are not independent like those of classical Brownian motion. To see this, see the covariance of two non-overlapping increments

$$E\left(\Delta B_{t,s}^H \Delta B_{s,0}^H\right) = E\left((B_t^H - B_s^H)(B_s^H - B_0^H)\right)$$
$$= E(B_t^H B_s^H) - E(B_t^H B_0^H) - E((B_s^H)^2) + E(B_s^H B_0^H)$$
$$= \frac{1}{2}\left[t^{2H} + s^{2H} - (t-s)^{2H}\right] - s^{2H}$$
$$= \frac{1}{2}\left[t^{2H} - s^{2H} - (t-s)^{2H}\right].$$

Consequently, apart from the special case $H = \frac{1}{2}$, the increments of fractional Brownian motion are no longer independent, but are correlated. The degree and the characteristics of this dependence will be examined in the following section.

Another important property, which the general fractional Brownian motion adopts from the special case of classical Brownian motion, is that of self-similarity. Roughly speaking, this property describes the fact that no matter which level of scale is chosen the process qualitatively looks the same. Recall that this is just the idea that one associates when thinking of fractals. More precisely, a stochastic process X_t is called self-similar with parameter a, if for any constant $c > 0$, the processes X_{ct} and $c^a X_t$ are identical in distribution. To put it in another way, compressing or uncompressing the process by a factor only changes the characteristics of the process up to a scaling of the axis of ordinates.

Fractional Brownian motion processes are self-similar with parameter H, that is, the Hurst parameter is also the self-similarity parameter (see Mandelbrot and van Ness (1968)). In Fig. 2.6, the relatively flat, grey line depicts the original process B_t^H, whereas the darkest line is the compressed version of B_{10t}^H and the third line is the scaled process $10^H B_t^H$. The similarities of the

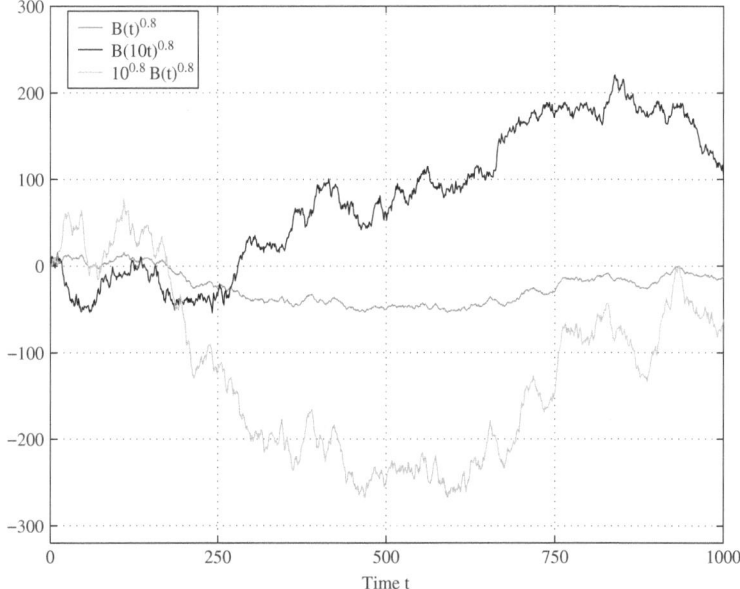

Fig. 2.6 Scaling properties of fractional Brownian motion (chosen parameter $H = 0.8$)

scaled and the compressed process are evident.

We have so far stated that fractional Brownian motion is a Gaussian process with self-similar and stationary increments. We now take a look at the role of the Hurst parameter H.

2.2 Serial Correlation: The Role of the Hurst Parameter

We mentioned above that the different ranges of possible Hurst parameters divide the family of fractional Brownian motion into three groups that can be distinguished by typical criteria. In particular, we will call processes with Hurst parameter H smaller than one half antipersistent, those with $H > \frac{1}{2}$ are called persistent. To get a first idea with regard to an explanation for these labels, take a look at the fractional increment

$$\Delta B^H(t) = B^H_{t+\Delta t} - B^H_t$$

$$= c_H \int_t^{t+\Delta t} (t + \Delta t - s)^{H-\frac{1}{2}} \, dB_s$$

$$+ c_H \int_{-\infty}^t \left[(t + \Delta t - s)^{H-\frac{1}{2}} - (t - s)^{H-\frac{1}{2}} \right] dB_s.$$

The first term contains the current innovation or shock, positively weighted for any parameter $0 < H < 1$. The second part of the sum is a moving average over all historical shocks, however the sign of the weights depends on the Hurst parameter: The weighting kernel is nothing else but an increment of the function $f(x) = x^{H-\frac{1}{2}}$. For $H < \frac{1}{2}$ this function is a hyperbola and downward-sloping, yielding negative increments. Aside from the random innovation in time t, the influence of the fractional increment for $H < \frac{1}{2}$ always results in a reversal of the past evolution. On the other hand, in the case $\frac{1}{2} < H < 1$, the weighting kernel is a radical function yielding positive weights. Therefore, the fractional Brownian increment positively depends on the generating Brownian motion, or more precisely on its historical increments. Clearly, in the case $H = \frac{1}{2}$, there is no influence from the historical shocks whatsoever, as the classical Brownian increment is independent of the past.

The increments of the different processes can also be characterized using the according autocovariance properties. We briefly recall some definitions. A stochastic process has short memory provided its autocovariance function declines at least exponentially when the lags are increased. Intermediate memory exists whenever its autocovariance function only declines hyperbolically but the infinite sum of all absolute values of autocovariances still exists. If the latter condition is no longer fulfilled, one speaks of long memory (see e.g. Barth (1996)). This implies the following rule if one wants to predict part of the future by looking at the past: restricting observations to only the finite past is feasible to the intermediate memory processes. With long memory however, we do not use a finite past, as by definition, the influence of the whole history must be taken into consideration.

If we examine the autocovariance function of the stationary process of discrete increments ΔB^H, for example for increments of length 1 and if we denote the lag size with τ, we obtain the following result:

$$\begin{aligned}
\gamma_H(\tau) = \ & E\left[(B_H(t+\tau) - B_H(t+\tau-1))(B_H(t) - B_H(t-1))\right] \\
= \ & E\left[B_H(t+\tau)B_H(t)\right] - E\left[B_H(t+\tau)B_H(t-1)\right] \\
& -E\left[B_H(t+\tau-1)B_H(t)\right] + E\left[B_H(t+\tau-1)B_H(t-1)\right] \\
= \ & \frac{1}{2}\left[(t+\tau)^{2H} + t^{2H} - \tau^{2H} - (t-1)^{2H} + (\tau+1)^{2H} - (t+\tau-1)^{2H}\right. \\
& \quad - t^{2H} - (t+\tau-1)^{2H} - t^{2H} + (\tau-1)^{2H} + (t+\tau-1)^{2H} \\
& \quad \left. + (t-1)^{2H} - \tau^{2H}\right] \\
= \ & \frac{1}{2}\left[(\tau+1)^{2H} - 2\tau^{2H} + (\tau-1)^{2H}\right].
\end{aligned}$$

Note that this term is also an approximation of the second derivative of the function $f(\tau) = \tau^{2H}$, it is the so-called central finite difference. In other words, for large τ, the autocovariance function behaves like the second derivative

$$f''(\tau) = 2H(2H-1)\tau^{2H-2}.$$

Recalling some basics about the theory of infinite sums, it is easy to verify that the infinite sum of values of the latter function only exists for exponents smaller than minus one. Accordingly, the sum exists for $H < \frac{1}{2}$ but is unlimited for the case $H > \frac{1}{2}$. Using the terms above, we state that the antipersistent fractional Brownian motion has intermediate memory, whereas the persistent one has long memory.

These results can be further illustrated by plotting the autocovariance functions for different Hurst parameters (see also Barth (1996), p. 55–57). Figure 2.7 depicts the autocovariance function for lags between zero and ten units of time. The selected Hurst parameters are larger than one half, that is we first focus on persistence. All the curves are bounded by an upper and a lower limiting curve. The upper boundary is the flat line of total persistence, where the autocovariance between increments of any lag would equal one. If one knows just one single realization of the process, the complete process can be extrapolated, as no additional randomness remains. The lower boundary for the persistent processes is the case of serial independence, where for overlapping intervals (in this case: lags smaller than one unit of time) the autocovariance equals the overlapping portion. Similarly, for non-overlapping intervals the autocovariance is zero. Within these limits, a higher degree of persistence implies a curve declining more slowly.

A depiction of the antipersistent case is illustrated by Fig. 2.8. Again, there are two situations of particular significance: The already well-known serial independent classical Brownian motion, on the one hand, and the limiting case of an absolutely antipersistent process, on the other hand. The latter would yield an autocovariance function that differs from zero in only two

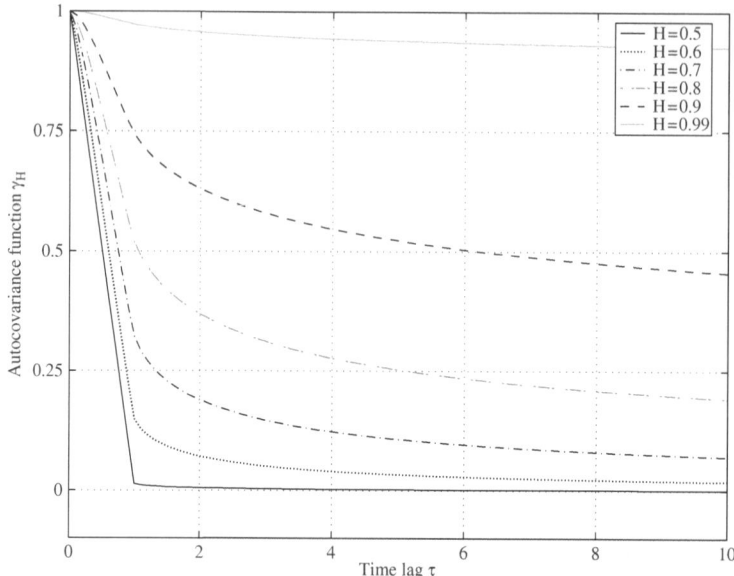

Fig. 2.7 Autocovariance function of fractional Brownian motion for the case of persistence (chosen parameters $H = 0.5$, $H = 0.6$, $H = 0.7$, $H = 0.8$, $H = 0.9$, $H = 0.99$)

cases, either when there is a total overlap between the two increments or in the case of two neighboring increments of equal length. It therefore can be interpreted as a linear combination of two Dirac–Delta functions. In Fig. 2.8, where the reference intervals have a length of one unit of time, these two cases occur when there is a zero lag or when the lag is equal to one unit of time. For example, let the reference increment be $B^H(t) - B^H(t - 1)$: Aside from the identical increment, which yields the variance of one, there are only two further correlated increments, which are $B^H(t+1) - B^H(t)$ and $B^H(t - 1) - B^H(t - 2)$. These are the bordering intervals of same length, where the covariance has the negative value of minus one half. To put it a different way, in the case of total antipersistence, it is sufficient to know the last increment of the process to predict the next increment. The future is negatively correlated with this immediate past.

For the Hurst parameters symbolizing antipersistence and lying between zero and one half, the shape of the autocovariance function can be described as follows: The section of overlapping increments with lags smaller than one all start with the unit variance. They then show a declining covariance until the situation where both increments are next to each other. Moreover we observe: The lower the degree of antipersistence, the lower the absolute value of covariance for $\tau = 1$ and the smoother the course of the curve between zero

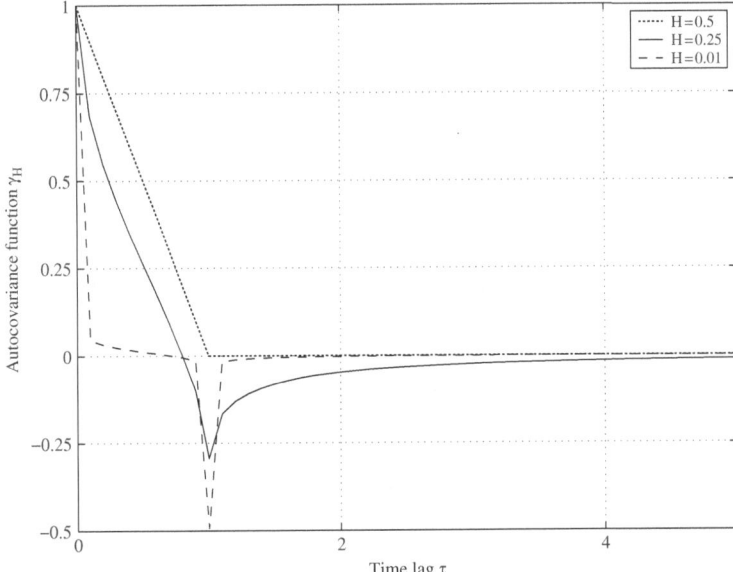

Fig. 2.8 Autocovariance function of fractional Brownian motion for the case of antipersistence (chosen parameters $H = 0.5$, $H = 0.25$, $H = 0.01$)

and one. For the non-overlapping increments ($\tau > 1$) all curves tend towards zero as distance grows. In fact, with an increasing level of antipersistence, the curves converge towards zero at a quicker rate.

Throughout this work we will focus on both the persistent case and the antipersistent case, however, occasionally we will draw comparisons to the classical Brownian theory.

2.3 The Wick-Based Approach to Fractional Integration

In the beginning of this chapter, we introduced the definition of a stochastic integral with respect to fractional Brownian motion when the integrand is deterministic. We required this integrand to be approximated by simple functions being piecewise constant and multiplied the respective function value with the discrete fractional Brownian increment. These Riemann sums converge in the mean square sense to a random variable, which we define to be the fractional integral. However, if the integrand is no longer deterministic and for example of infinite variation, convergence in the mean square sense is not necessarily given (see e.g. Gripenberg and Norros (1996), p. 4). In principle, there are two ways out of this issue. The first one is to modify the way the Riemann sums are built and/or to change (and perhaps relax) the definition of convergence. We will focus on this possibility in the next section. A very elegant solution following the first idea, is mainly due to Duncan et al. (2000), which we will refer to from time to time. The main aspect of their approach is easy to grasp: Replace ordinary multiplication within the Riemann sums with a different multiplicative concept, called the Wick product which is denoted by the so-called diamond symbol $\diamond$. However, if we want to understand the characteristics of this Wick product—which will perhaps seem a little bit peculiar at first glance—we need to introduce some mathematical notation as well as some basic results of fractional white noise calculus (see e.g. Hu and Øksendal (2003)).

Motivated by the already-stated results by Gripenberg and Norros (1996) concerning fractional integrals with deterministic integrands, we introduce the fractional kernel φ as a function $\varphi : \mathbb{R}^2 \to \mathbb{R}$ defined by

$$\varphi(s,t) = H(2H-1)|s-t|^{2H-2}.$$

Furthermore we endow the space of deterministic functions $f : \mathbb{R} \to \mathbb{R}$ with the norm $|\cdot|_\varphi^2$:

$$|f|_\varphi^2 := \int_0^\infty \int_0^\infty f(s)f(t)\varphi(s,t)\,ds\,dt.$$

Actually, for the case $H = \frac{1}{2}$, the fractional kernel φ is the Dirac–Delta function and we obtain

$$|f|_{\frac{1}{2}}^2 := \int_0^\infty f(t)^2\,dt.$$

For all other Hurst parameters, this norm can be interpreted as a blurred version of the well-known norm of square integrability $|.|^2$, integrating not only over the bisector of $\mathbb{R}^2$ but also factoring neighboring products into the integration. Accordingly, the inner product $\langle \cdot, \cdot \rangle_\varphi$ of two functions f and g within the resulting Hilbert space L_φ^2 is

$$\langle f,g \rangle_\varphi := \int_0^\infty \int_0^\infty f(s)g(t)\varphi(s,t)\,ds\,dt.$$

Using this, we can rewrite (2.3) as follows:

$$E\left(\int_0^\infty f(s)\,dB_s^H \int_0^\infty g(t)\,dB_t^H \right) = \langle f,g \rangle_\varphi.$$

Since our goal is to generalize the integration concept to stochastic integrands, we first introduce the probability space $(\Omega, \mathfrak{F}, P)$, where Ω is the set of possible states, $\mathfrak{F}$ is a σ-algebra on Ω and P is the probability measure. We accept in the following all random variables X as integrands which are defined on this probability space and which satisfy

$$(E|X|^p)^{\frac{1}{p}} < \infty.$$

We denote the space of these p-integrable random variables by L^p. Duncan et al. (2000) show that for any $p \geq 0$, the random variables of L^p can be approximated with arbitrary exactness by linear combinations of so-called Wick exponentials $\varepsilon(f)$ that are defined via fractional integrals with deterministic integrands f:

$$\varepsilon(f) := \exp\left(\int_0^\infty f(t)dB_t^H - \frac{1}{2}|f|_\varphi^2\right), \quad f \in L_\varphi^2.$$

Note that such an exponential is of course a random variable. As the stochastic integral $\int_0^\infty f(t)\,dB_t^H$ is normally distributed with zero mean and variance $|f|_\varphi^2$ (see Gripenberg and Norros (1996), p. 4), $\exp\left(\int_0^\infty f(t\,dB_t^H)\right)$ is log-normally distributed with mean $\exp\left(\frac{1}{2}|f|_\varphi^2\right)$. Hence we get

$$E\left(\varepsilon(f)\right) = E\left(\exp\left(\int_0^\infty f(t\,dB_t^H)\right)\right)\exp\left(-\frac{1}{2}|f|_\varphi^2\right)$$

$$= \exp\left(\frac{1}{2}|f|_\varphi^2\right)\exp\left(-\frac{1}{2}|f|_\varphi^2\right) = 1.$$

Moreover, we have

$$\varepsilon(f)\varepsilon(g) = \exp\left(\int_0^\infty f(s)\,dB_s^H - \frac{1}{2}\int_0^\infty\int_0^\infty \varphi(u,v)f(u)f(v)\,dudv\right)$$

$$\times \exp\left(\int_0^\infty g(s)\,dB_s^H - \frac{1}{2}\int_0^\infty\int_0^\infty \varphi(u,v)g(u)g(v)\,dudv\right)$$

$$= \varepsilon(f+g)\exp\left(\int_0^\infty\int_0^\infty \varphi(u,v)f(u)g(v)\,dudv\right) \quad \cdot$$

$$= \varepsilon(f+g)\langle f,g\rangle_\varphi,$$

and therefore

$$E\left(\varepsilon(f)\varepsilon(g)\right) = \langle f,g\rangle_\varphi.$$

or

$$E\left(\varepsilon(f)^2\right) = |f|_\varphi^2,$$

respectively.

Duncan et al. (2000) now define the Wick product implicitly on these Wick exponentials by postulating that for two functions $f,g \in L_\varphi^2$ the following equation holds:

$$\varepsilon(f) \diamond \varepsilon(g) = \varepsilon(f + g).$$

From this definition on, the Wick product can be extended to random variables in L^p. For an explicit definition of the Wick product based on a representation using Hermite polynomials, see for example Hu and Øksendal (2003). We stress that the Wick product of two random variables is only defined as a multiplication of two complete random variables and cannot be interpreted in a pathwise sense. In particular, if one knows nothing but the realizations of two random variables F, G, that is $F(\omega)$ and $G(\omega)$, it will not be possible to calculate $(F \diamond G)(\omega)$. Another peculiarity that is worth mentioning stems from the same reason. The combination of ordinary products and Wick products is not as easy as one might suggest. Actually, despite each of them being associative on their own, it is not possible to change the order of multiplication when products of both types are involved; that is, in general, we have for $F, G, H \in L^p$ (see Bender (2003a), p. 13):

$$(F \diamond G) \cdot H \neq F \diamond (G \cdot H).$$

This non-compatibility will be focused on later when introducing the fractional market setup.

As an immediate consequence of the above definitions and calculations, we obtain the moment properties of a Wick product of two wick exponentials:

$$E\left(\varepsilon(f) \diamond \varepsilon(g)\right) = E\left(\varepsilon(f + g)\right) = 1 = E\left(\varepsilon(f)\right) E\left(\varepsilon(g)\right)$$
$$E\left(\varepsilon(f) \diamond \varepsilon(g)\right)^2 = E\left((\varepsilon(f + g))^2\right) = |f + g|_\varphi^2.$$

These results can be extended to Wick products of the form $\varepsilon(f) \diamond \int_0^\infty g(t) \, dB_t^H$ (see Duncan et al. (2000), p. 588–590). They receive

$$E\left(\varepsilon(f) \diamond \int_0^\infty g(t) \, dB_t^H\right) = E\left(\varepsilon(f)\right) E\left(\int_0^\infty g(t) \, dB_t^H\right)$$
$$E\left(\varepsilon(f) \diamond \int_0^\infty g(t) \, dB_t^H\right)^2 = \exp(|f|_\varphi^2) \left((\langle f, g \rangle_\varphi)^2 + |g|_\varphi^2\right).$$

They further show that this again can be generalized to the case when the first factor is a random variable F in L^p (see also Holden et al. (1996), p. 83):

$$E\left(F \diamond \int_0^\infty g(t) \, dB_t^H\right) = E\left(F\right) E\left(\int_0^\infty g(t) \, dB_t^H\right)$$
$$E\left(F \diamond \int_0^\infty g(t) \, dB_t^H\right)^2 = E\left(\left(\int_0^\infty D^\varphi F_s \, ds\right)^2 + F^2 |g|_\varphi^2\right),$$

where $D^\varphi F$ denotes a version of the Malliavin derivative of F (for an easy approach to Malliavin calculus, see Øksendal (1996)).

If we now look at the Riemann sums $S_\diamond(F, \pi)$ of Wick type with respect to a partition π, that is

$$S_\diamond(F, \pi) := \sum_{i \in \pi} F(t_i) \diamond \left(B_{t_{i+1}}^H - B_{t_i}^H \right),$$

and if partitions become finer ($|\pi| \to 0$), one defines the fractional integral of Wick type as the limit of the according sequence of Riemann sums

$$\int_0^T F(s) \, dB_s^H := \lim_{|\pi| \to 0} \sum_{i \in \pi} F(t_i) \diamond \left(B_{t_{i+1}}^H - B_{t_i}^H \right).$$

Using some of the results above and introducing some regularity conditions (see Duncan et al. (2000), p. 591) the moment properties of the fractional integral of Wick type can be derived:

$$E \left(\int_0^T F(s) \, dB_s^H \right) = \lim_{|\pi| \to 0} E \left(\sum_{i \in \pi} F(t_i) \diamond \left(B_{t_{i+1}}^H - B_{t_i}^H \right) \right)$$

$$= \lim_{|\pi| \to 0} E \left(\sum_{i \in \pi} F(t_i) \diamond \int_{t_i}^{t_{i+1}} dB_s^H \right)$$

$$= \lim_{|\pi| \to 0} E \left(\sum_{i \in \pi} F(t_i) \right) E \left(B_{t_{i+1}}^H - B_{t_i}^H \right)$$

$$= 0$$

$$E \left(\int_0^T F(s) \, dB_s^H \right)^2 = \lim_{|\pi| \to 0} E \left(\sum_{i \in \pi} F(t_i) \diamond \int_{t_i}^{t_{i+1}} dB_s^H \right)$$

$$= E \left(\left(\int_0^T D^\varphi F_s \, ds \right)^2 + |F|_\varphi^2 \right).$$

Due to the evident parallels to classical Brownian integration theory, the latter property is called fractional Itô isometry. We further observe that the fractional integral of Wick type has zero mean, which of course is a convenient feature of the new integration theory. Actually, this is one of the main reasons why Wick integrals are introduced in the financial context. We will see in the next section that the integration concepts of pathwise and Stratonovich type may be more intuitive and easier to deal with but that they cannot provide this desirable zero-mean property.

2.4 Pathwise and Stratonovich Integrals

Instead of using the Wick product as we suggested in the section above, it is also possible to define an integration concept based on ordinary multiplication. However, in this case, the meaning of convergence has to be redefined. Actually, there are two alternative ways on how convergence can be relaxed from the mean square sense. Either one postulates pathwise convergence, that is for any state of nature, i.e. path, the integral—then being deterministic—converges in the Riemann–Stieltjes sense. The second alternative results if the approximating sums are required to tend to their limit in probability. Regarding the first concept, one has to ensure the pathwise Hölder continuity of the integrand (see Zähle (1998)) and due to its defining property it is called pathwise integration. The integral based on convergence in probability is called fractional integral of Stratonovich type and allows for a wider class of integrands. Recall that the classical Stratonovich integral with respect to Brownian motion differs from the well-known Itô integral as the integrand is always evaluated in the middle of an interval (see Stratonovich (1966)). In this case, the integral becomes anticipating, whereas Itô integrals exploit the left boundaries of the intervals. Lin (1995) introduced this integral for the case $H > \frac{1}{2}$ and chose for the Riemann sums the presentation

$$S(F, \pi) := \sum_{i \in \pi} F(t_i) \left(B^H_{t_{i+1}} - B^H_{t_i} \right).$$

This means that the value of the integrand is always taken at the earliest point of the interval. Surprisingly, this integral is nevertheless rather related to the anticipating classical Stratonovich integral that uses the midpoints of the intervals than to the classical Itô integral. Actually, it can be shown (see Duncan et al. (2000), p. 595), that for the persistent case the parametrization of the evaluation point does not matter. Yet, Bender (2003a) proves that if one wants to extend the concept to the antipersistent domain, the integral will converge if and only if the midpoint is chosen (see Bender (2003a), p. 79), so we define the fractional integral of Stratonovich type $\int_0^T F(s) \, \delta B^H_s$ in the following way:

$$\int_0^T F(s) \, \delta B^H_s := \lim_{|\pi| \to 0} \sum_{i \in \pi} F\left(\frac{t_{i+1} + t_i}{2} \right) \left(B^H_{t_{i+1}} - B^H_{t_i} \right).$$

As the pathwise fractional integral and the fractional integral of Stratonovich type share the most important properties, we consider—as done in most of the literature—both concepts from now on as one single approach to fractional integration using both terms equivalently.

Duncan et al. (2000) p. 592, proved that there is an easy link between the fractional integral of Wick–Itô type and that of Stratonovich type:

$$\int_0^T F(s)\,\delta B_s^H = \int_0^T F(s)\,dB_s^H + \int_0^T D^\varphi F_s\,ds.$$

The result includes again the term $D^\varphi F$ which is the fractional version of the Malliavin derivative of the integrand. To get an idea of this derivative, we recall a result by (Duncan et al. (2000) p. 588) concerning the fractional Malliavin derivative of a fractional integral with deterministic integrand $\int_0^T f(s)\,dB_s^H$:

$$D^\varphi \left(\int_0^T f(u)\,dB_u^H \right)(s) = \int_0^T \varphi(u,s)f(u)\,du.$$

For example, the fractional Malliavin derivative of B_s^H in time s is

$$D^\varphi \left(B_s^H \right)(s) = D^\varphi \left(\int_0^s dB_u^H \right)(s)$$

$$= \int_0^s \varphi(u,s)\,du$$

$$= \int_0^s H(2H-1)|u-s|^{2H-2}\,du$$

$$= H|-s|^{2H-1}.$$

For the special case $H = \frac{1}{2}$, we hence get the classical Malliavin derivative

$$D \left(B_s^H \right)(s) := lim_{H \to \frac{1}{2}} D^\varphi \left(\int_0^s dB_u^H \right)(s)$$

$$= lim_{H \to \frac{1}{2}} \left(H|-s|^{2H-1} \right)$$

$$= \frac{1}{2}.$$

Tables 2.1 and 2.2 summarize the different properties of the Itô integrations approaches on the one hand and those of the integrals of Stratonovich type on the other hand. Furthermore, the comparison with the case $H = \frac{1}{2}$ illustrates the parallels of the concepts.

The main reason why one prefers Itô integrals in the classical financial context, is the non-anticipating character of the integral. To put it differently, the anticipating property of the Stratonovich integral makes it less applicable for a financial setting where one normally cannot predict the future. For the case of integration with respect to classical Brownian motion, Sethi and Lehoczky (1981) derived a very illustrative example, where this can lead to. More precisely, they reformulated the Black–Scholes setting using Stratonovich integration calculus and derived a formula for a European call option where no randomness is left and the value of the option $C(t, S_t)$ with strike K and

$$H = \tfrac{1}{2}$$

Classical Itô Integral	**Classical Stratonovich Integral**
$\int_0^T F(s)\,dB_s$	$\int_0^T F(s)\,\delta B_s$
$=$	$=$
$\lim \sum F(t_i)(B_{t_{i+1}} - B_{t_i})$	$\lim \sum F\left(\frac{t_{i+1}+t_i}{2}\right)(B_{t_{i+1}} - B_{(t_i)}))$
limit of left point Riemann sums using ordinary multiplication	limit of midpoint Riemann sums using ordinary multiplication
martingale property	no martingale property

Interrelation Between the Two Integrals

$$\int_0^T F(s)\,dB_s = \int_0^T F(s)\,\delta B_s - \int_0^T DF_s\,ds$$

Example:

$$\int_0^T B(s)\,dB_s = \tfrac{1}{2}B_T^2 - \tfrac{1}{2}T = \int_0^T B(s)\,\delta B_s - \int_0^T DB_s\,ds$$

Table 2.1 Comparison and relations between integrals of Itô type and of Stratonovich type

maturity T reduces to

$$C(t, S_t) = \max(S_t - Ke^{-r(T-t)}, 0).$$

The authors show however, that the problem can be solved, if—for reasons of plausibility—the weighted stock position in the partial differential equation is interpreted as an Itô differential, that has to be transformed to the Stratonovich notation. In Sect. 4.3 we will reconsider these results by Sethi and Lehoczky (1981) and extend them to the fractional context. The results there will look quite astonishing; moreover, some of the arguments used in the classical case will be totally inverted and we will see that both integrals yield some kind of predictability.

Within our fractional context, we have to state: As the integrator of fractional Brownian motion is not a semi-martingale, neither the Stratonovich nor the Wick–Itô integral can generate an integral that exhibits martingale properties. Actually, this will be the reason why we will have to give up no arbitrage valuation, because the available information now tells something about the shape of future distribution. In combination with the possibility to interact infinitely fast, these predictions can be exploited. Nevertheless, it

$$H \neq \tfrac{1}{2}$$

Fractional Wick–Itô Integral   **Fractional Stratonovich Integral**

$$\int_0^T F(s)\, dB_s^H$$
$$=$$
$$\lim \sum F(t_i) \diamond \left(B_{t_{i+1}}^H - B_{t_i}^H \right)$$

$$\int_0^T F(s)\, \delta B_s^H$$
$$=$$
$$\lim \sum F\left(\tfrac{t_{i+1}+t_i}{2} \right) \left(B_{t_{i+1}}^H - B_{t_i}^H \right)$$

limit of left point Riemann sums   limit of midpoint Riemann sums
using Wick multiplication   using ordinary multiplication

integral has zero mean   integral has mean $\neq 0$

Interrelation Between the Two Integrals

$$\int_0^T F(s)\, dB_s^H = \int_0^T F(s)\, \delta B_s^H - \int_0^T D^\varphi F_s\, ds$$

Example:

$$\int_0^T B^H(s)\, dB_s^H = \tfrac{1}{2}(B_T^H)^2 - \tfrac{1}{2}T^{2H} = \int_0^T B^H(s)\, \delta B_s^H - \int_0^T D^\varphi B_s^H\, ds$$

Table 2.2 Comparison and relations between integrals of Wick–Itô type and of fractional Stratonovich type

seems to be more plausible to have an integral that has at least zero mean, that means, over all possible paths there occurs no systemic bias in the price process. Another argument for the Wick–Itô approach is the formal compatibility with classical Brownian theory which will turn out in all its beauty in Chap. 5 when we price options.

2.5 Some Important Results of the Wick Type Fractional Integration Calculus

In the foregoing section we elaborated the advantages of the integration theory based on the Wick product. We have already stated some first parallels to the classical Itô calculus. In this section we present further important results, in particular, analogons to the central theorems of the Itô theory are given. By name, we state fractional versions of the Girsanov theorem and the Itô formula. We only provide the results and skip all proofs referring to the relevant literature. For a summarizing discussion of the topic, see Bender (2003a).

The classical Girsanov theorem discusses the properties of classical Brownian motion—or more generally classical Brownian integrals—under change of measure. It gives the possibility of changing a Brownian motion with drift into one without any drift. The same is possible in the fractional context. Norros and Valkeila (1999) p. 13 et seq., proved that, if X_t is a fractional Brownian motion with drift under the measure P, that is

$$X_t \cong B_t^H + at \quad \text{under} \quad P, \tag{2.4}$$

then, there is a suitable measure P^a so that

$$X_t \cong B_t^H \quad \text{under} \quad P^a,$$

that is, X_t is a fractional Brownian motion without drift under the new measure.

The change of measure is given via the Radon–Nikodym derivative

$$\frac{dP^a}{dP} = \exp\left(-aM_t - \frac{1}{2}a^2 \langle M, M \rangle_t\right), \tag{2.5}$$

where

$$M_t = \int_0^t c_1 s^{\frac{1}{2}-H}(t-s)^{\frac{1}{2}-H} dB_s^H,$$

$$\text{and} \quad c_1 = \left[2H\Gamma\left(\frac{3}{2}-H\right)\Gamma\left(H+\frac{1}{2}\right)\right]^{-1}.$$

The process M_t is a martingale with independent increments, zero mean and variance function

$$EM_t^2 = c_2^2 t^{2-2H},$$

$$\text{where} \quad c_2 = \frac{c_H}{2H\sqrt{2-2H}}.$$

It is called the fundamental martingale. Using this, we can rewrite (2.5) of the Radon–Nikodym derivative by

$$\frac{dP^a}{dP} = \exp\left(-aM_t - \frac{1}{2}a^2 c_2^2 t^{2-2H}\right).$$

From this representation it is easy to see that for the case $H = \frac{1}{2}$ we obtain the well-known change of measure formula. The generalization of this drift removal theorem from the simple fractional Brownian motion to fractional integrals can also be done (see Bender (2003b), p. 973 et seq.). However, to provide this, the notations and concepts of the more powerful S-transform are needed, so we postpone the result to Sect. 2.6.

The second outstanding result of the fractional integration calculus of Wick type is the fractional Itô theorem, that is, a chain rule for processes based on fractional Brownian motion. Recall that for the case of classical Brownian motion, we have different chain rules, depending on which integral definition we choose. In the case of Stratonovich integrals, the chain rule is identical to the deterministic case. However, when choosing Itô integrals, a correction terms occurs including the second derivative of the outer function with respect to the random process. In fractional calculus, the parallels again are highly visible. For the fractional Stratonovich integral, the chain rule still resembles its deterministic origin. In contrast, the fractional Wick approach also necessitates a correcting second derivative term. In its basic version, the fractional Itô theorem is due to Duncan et al. (2000). Let F, G be stochastic processes satisfying certain regularity conditions. For the stochastic process

$$X_t = X_0 + \int_0^t G_u \, du + \int_0^t F_u \, dB_u^H,$$

and a twice continuously differentiable function $f : \mathbb{R}^2 \to \mathbb{R}$ the process $f(t, X_t)$ satisfies the following equation:

$$f(t, X_t) = f(0, X_0) + \int_0^t \frac{\partial f}{\partial s}(s, X_s) \, ds + \int_0^t \frac{\partial f}{\partial x}(s, X_s) \, ds$$
$$+ \int_0^t \frac{\partial f}{\partial x}(s, X_s) F_s \, dB_s^H + \int_0^t \frac{\partial^2 f}{\partial x^2}(s, X_s) F_s D^\varphi X_s \, ds. \quad (2.6)$$

As a special case, we obtain for $X_t = B_t^H$ and a not time-dependent function $g : \mathbb{R} \to \mathbb{R}$

$$f(B_t^H) = f(B_0^H) + \int_0^t f'(B_s^H) \, dB_s^H + H \int_0^t s^{2H-1} f''(B_s^H) \, ds. \quad (2.7)$$

Once more, it can be easily verified that the limit case $H = \frac{1}{2}$ yields the well-known Itô formula.

As already mentioned above, it is important to stress the fact that the derivation of these formulae is only valid for the case of persistence but not for $H < \frac{1}{2}$. Again, like for the Girsanov formula, the more general results for all Hurst parameters rely on the S-transform approach that we will briefly present in the following section.

2.6 The S-Transform Approach

The presented approaches to fractional integration theory whether of Itô or Stratonovich type all had one property in common: In the case of Hurst parameters smaller than one half—that is, as antipersistence occurs—there are severe problems concerning the convergence of the Riemann sums. By means of the S-transform approach introduced by Bender (2003b), both Wick–Itô type and Stratonovich type integration can be extended to antipersistent processes. In the following, we briefly present the basic idea of this concept.

First, we have to introduce some notation. For $\frac{1}{2} < H < 1$, the Riemann–Liouville fractional integrals are defined by

$$I_-^{H-\frac{1}{2}} f(x) = \frac{1}{\Gamma(H - \frac{1}{2})} \int_x^\infty f(s)(s - x)^{H-\frac{3}{2}} \, ds,$$

$$I_+^{H-\frac{1}{2}} f(x) = \frac{1}{\Gamma(H - \frac{1}{2})} \int_{-\infty}^x f(s)(x - s)^{H-\frac{3}{2}} \, ds.$$

These fractional integrals are nothing but normalized blurred versions of the function f, either averaging over future or over past function values. For $H > \frac{1}{2}$, the weights become smaller the greater the distance to the proper argument of the function gets. On the other hand, the fractional derivative of Marchaud's type $D_\pm^{-(H-\frac{1}{2})}$ of a function f is given by

$$D_\pm^{-(H-\frac{1}{2})} f := \lim_{\epsilon \to 0} -\frac{H - \frac{1}{2}}{\Gamma(H - \frac{1}{2})} \int_\epsilon^\infty \frac{f(x) - f(x \mp t)}{t^{\frac{3}{2}-H}} \, dt$$

$$= \lim_{\epsilon \to 0} -\frac{H - \frac{1}{2}}{\Gamma(H - \frac{1}{2})} \int_\epsilon^\infty \frac{f(x) - f(x \mp t)}{t} t^{H-\frac{1}{2}} \, dt.$$

Concerning the latter representation, we can also interpret this fractional derivative as a weighted sum, this time averaging difference quotients, yielding a blurred version of the first derivative of f. Based on these definitions, the operators $M_\pm^H$ are defined by

$$M_\pm^H f := \begin{cases} K_H D_\pm^{-(H-\frac{1}{2})} f & 0 < H < \frac{1}{2}, \\ f & H = \frac{1}{2}, \\ K_H I_\pm^{H-\frac{1}{2}} f & \frac{1}{2} < H < 1, \end{cases}$$

where

$$K_H = \Gamma\left(H + \frac{1}{2}\right) \sqrt{\frac{2H\Gamma(\frac{3}{2} - H)}{\Gamma(H + \frac{1}{2})\Gamma(2 - 2H)}}.$$

Note that the process of fractional Brownian motion can be represented using this operator. Like the original representation by Mandelbrot and van Ness (1968), it is a stochastic integral with respect to classical Brownian motion. In particular, we have

$$B_t^H = \int_{\mathbb{R}} \left(M_-^H \mathbf{1}_{[0,t]} \right)(s) \, dB_s. \tag{2.8}$$

The identity of these two representations is carried out in Bender (2003c). Using the operators $M_{\pm}^H$, one can formulate the useful fractional integration by parts rule (see Bender (2003b), p. 960):

$$\int_{\mathbb{R}} f(s) \left(M_-^H g \right)(s) \, ds = \int_{\mathbb{R}} \left(M_+^H f \right)(s) g(s) \, ds. \tag{2.9}$$

We now introduce the S-transform. The S-transform of a mean square integrable random variable F is the functional SF operating on deterministic functions g and is fully characterized by the following defining equation:

$$SF(g) := E\left(F \exp\left(\int_{\mathbb{R}} g(s) \, dB_s - \frac{1}{2}|g|^2 \right) \right).$$

For example, (Bender (2003b), p. 964) shows that the S-transform of a simple Wiener integral $\int_a^b f(t) \, dB_t$ is

$$S\left(\int_a^b f(t) \, dB_t \right)(g) = \int_a^b f(t) g(t) \, dt. \tag{2.10}$$

From there, we obtain as another easy result

$$S\left(B_t^H \right)(g) = S\left(\int_0^t f(s) \, dB_s \right)(g) = \int_0^t g(s) \, ds.$$

Furthermore, the S-transform of the classical Itô integral $\int_a^b X_t \, dB_t$ satisfies

$$S\left(\int_a^b X_t \, dB_t \right)(g) = \int_a^b (SX_t)(g) g(t) \, dt = \int_a^b S(X_t)(g) \frac{d}{dt} S(B_t)(g) \, dt.$$

As the S-transform is injective, it also can be taken to define the above integrals. Drawing the conclusion by analogy, one can accordingly define the fractional integral of Wick–Itô type $\int_a^b X_t \, dB_t^H$ to be the unique random variable with S-transform

$$S\left(\int_a^b X_t \, dB_t^H \right)(g) = \int_a^b S(X_t)(g) \frac{d}{dt} S(B_t^H)(g) \, dt. \tag{2.11}$$

If we recall (2.8) and apply the S-transform on the Wiener integral following (2.10) as well as the fractional integration by parts rule (2.9), we receive

$$\frac{d}{dt} S(B_t^H)(g) = \frac{d}{dt} S\left(\int_{\mathbb{R}} \left(M_-^H \mathbf{1}_{[0,t]}\right)(s)\, dB_s\right)(g)$$

$$= \frac{d}{dt} \int_{\mathbb{R}} M_-^H \mathbf{1}_{[0,t]}(s) g(s)\, ds$$

$$= \frac{d}{dt} \int_0^t (M_+^H g)(s)\, ds$$

$$= (M_+^H g)(t).$$

Hence, (2.11) can be reformulated and the fractional Itô integral is defined to be the unique random variable with the following S-transform:

$$S\left(\int_a^b X_t\, dB_t^H\right)(g) = \int_a^b S(X_t)(g)(M_+^H g)(t)\, dt.$$

It can be shown (see Bender (2003a)) that this definition includes the Wick–Riemann sum approach by Duncan et al. (2000) and allows in addition for an extension to Hurst parameters smaller than one half. Based on this integral definition, we are able to generalize the results of Girsanov type and the fractional Itô formula. In comparison with (2.4) and (2.5), a fractional integral with drift can—after a suitable change of measure—be rewritten as one without drift. More precisely, if a random variable Y under the probability measure P can be written as

$$Y = \int_{\mathbb{R}} X_t\, dB_t^H + \int_{\mathbb{R}} X_t \left(M_+^H f\right)(t)\, dt,$$

then, applying the change of measure with the Radon–Nikodym derivative

$$\frac{dP^f}{dP} = \exp\left(\int_{\mathbb{R}} f(s)\, dB_s - \frac{1}{2}|f|^2\right),$$

Y can be represented as

$$Y = \int_{\mathbb{R}} X_t\, d\tilde{B}_t^H,$$

where $\tilde{B}_t^H$ again is a fractional Brownian motion, but now under the new measure P_f.

Concerning the extensions of the fractional Itô formula that are possible due to the S-transform approach, we stress one more time that the (2.6) and (2.7) now can also be proven for the case $H < \frac{1}{2}$ (see Bender (2003b), p. 976–979). At the same place, alternative representations are given using the operators

$M_\pm^H$. We pass on the formal specification and skip the details at this point in time, but we will resume the S-transform later on and then sketch and seize the basic steps of the proof, when we will need a conditional version of the fractional Itô theorem.

As a final remark regarding the S-transform approach, we mention that the concept of the S-transform can also be used in order to extend the fractional integral of Stratonovich type to all values of the Hurst parameter. Roughly speaking, for a sequence of partitions, (Bender (2003a), p. 72 et seq.) defines the Stratonovich type Riemann sums to converge to a random variable X if the respective S-transforms converge. The limit random variable is then called the fractional integral of Stratonovich type.

Although it is important from a technical and substantiating point of view, we will eclipse the S-transform approach within the following chapters and sections as far as possible. Instead, we will concern ourselves with the more intuitive Riemann sum approaches. With respect to the latter, the discrete time considerations in the next chapter will deliver deeper insight.

Chapter 3
Fractional Binomial Trees

Binomial trees are discrete approximations of stochastic processes where at every discrete point in time the process has two possibilities: it either moves upwards or descends to a certain extent. Each alternative occurs with a certain probability adding up to 1. Consequently, two factors determine the characteristics of the resulting discrete process: The probability distributions of the single steps as well as the extent of the two possible shifts at each step. The binomial tree approach for classical Brownian motion is well-developed and leads to intuitive insights concerning the understanding of Brownian motion as the limit of an uncorrelated random walk. Cox et al. (1979) extended the very setting and defined a binomial stock price model converging weakly to the lognormal diffusion of geometric Brownian motion. Other processes of several important continuous time models in finance have been modeled successfully in a similar fashion (see e.g. Nelson and Ramaswamy (1990)). Hence, one might expect that a comparable approach for fractional Brownian motion should also be possible. However, as we will see in this chapter, things are a little bit harder to work out. This is mainly due to the property of serial correlation.

We recall that fractional Brownian motion B_t^H with Hurst parameter H can be regarded as a moving average of a two-sided classical Brownian motion B_s (see Mandelbrot and van Ness (1968)):

$$B_t^H = c_H \left[\int_{\mathbb{R}} \left((t-s)_+^{H-\frac{1}{2}} - (-s)_+^{H-\frac{1}{2}} \right) dB_s \right], \qquad (3.1)$$

where c_H is a normalizing constant.

The chapter proceeds in the following way: In Sect. 3.1 we briefly present two different approaches of binomially approximating fractional Brownian motion. We provide a deeper discussion of one of them, the finite memory model by Sottinen (2001). In the subsequent chapter, we show in Sect. 3.2,

S. Rostek, *Option Pricing in Fractional Brownian Markets.*
Lecture Notes in Economics and Mathematical Systems.
© Springer-Verlag Berlin Heidelberg 2009

how conditional moments can be recovered and calculated within the binomial model by Sottinen (2001). In the fourth section of the chapter we will broach the issue of modelling a multiplicative price process comparable to that of Cox et al. (1979). In particular, we will show two ways of how to form binomial models of geometric fractional Brownian motion. We proceed with some remarks concerning arbitrage possibilities within the binomial setting. In particular, we suggest a solution to the problem of pricing options in the fractional binomial market. Finally, we conclude the chapter with a basic example illustrating the main idea of this approach.

3.1 Binomial Approximation of an Arithmetic Fractional Brownian Motion Process

By means of the central limit theorem, it is possible to approximate classical Brownian motion by increasing the number of independently and identically distributed random variables. Therefore, taking a binomial distribution for these random variables, one receives the binomial tree for classical Brownian motion (see Cox et al. (1979)). Moreover, there is no problem in extending this construction to a two-sided Brownian motion: As the two sides are not at all correlated, starting from 0, one can symmetrically evolve both sides of the process and receives a process over the whole time line.

Looking at (3.1), it is obvious that fractional Brownian motion also relates to an infinite past. Unfortunately, it is not promising to carry over the above idea of a two-sided symmetric approximation if one wants to model fractional Brownian motion by a binomial tree. The main reason for this is that for any time t the moving average of the historic realizations has to be recalculated completely, as the according weights depend on present time and therefore change. So, in order to evolve B_t^H, it is always necessary to factor the whole history—also beyond null—into the calculation, hence an independent symmetric modeling is inadequate.

Dasgupta (1998) suggests a way out that is able to incorporate the influence of an infinite past. For $t \in [0,1]$, the nth approximation of B_t^H is defined to be a weighted sum of $2^n + n + 1$ independently and identically binomial random variables $\xi_i^{(n)}$ taking values ± 1 with probability $\frac{1}{2}$ each:

$$B_t^{H(n)} = \sum_{i=-2^n}^{n} \left(\sqrt{n} \int_{\frac{i-1}{n}}^{\frac{i}{n}} f(t,s)\, ds \right) \xi_i^{(n)}, \qquad (3.2)$$

where

$$f(t,s) = c_H \left((t-s)_+^{H-\frac{1}{2}} - (-s)_+^{H-\frac{1}{2}} \right) \tag{3.3}$$

is the weighting kernel of fractional Brownian motion as in (2.2).

Dasgupta (1998) proves that this process weakly converges to fractional Brownian motion as n gets larger. However, there is one crucial drawback within this approximation procedure. The effort of calculating appropriate approximations is extremely high, as we obtain 2^{2^n+n+1} terminal values within the nth step. Hence, for practical purposes this approach is not applicable.

Obviously, there are some problems if one wants to model a process going infinitely back to the past, in particular, when any future step depends on the whole history. Instead, it is much more promising to model a process that starts at a fixed point in time $t = 0$. Therefore, we recall a representation of fractional Brownian motion as a finite Brownian integral which was given by Norros and Valkeila (1999). They derive a finite interval representation of fractional Brownian motion which reads as follows:

$$B_t^H = \int_0^t z(t,s)\, dB_s, \tag{3.4}$$

where

$$z(t,s) = c_H \left[\frac{t}{s}^{H-\frac{1}{2}} (t-s)^{H-\frac{1}{2}} - \left(H - \frac{1}{2} \right) s^{\frac{1}{2}-H} \int_s^t u^{H-\frac{3}{2}} (u-s)^{H-\frac{1}{2}}\, du \right].$$

For the proof, we refer to Norros and Valkeila (1999). With this latter equation being available, it is easy to grasp the idea of Sottinen (2001). As the fractional Brownian motion is now a weighted sum of Brownian increments going finitely back to the past, he approximates the standard Brownian motion by a sum of discrete independently and identically distributed random variables $\xi_i^{(n)}$ with zero mean and unit variance, that is

$$B_t^{(n)} := \frac{1}{\sqrt{n}} \sum_{i=1}^{[nt]} \xi_i(n), \tag{3.5}$$

where in the n-th approximation step each unit time interval is divided into n discrete steps and time t is rounded down onto the next n-th part by replacing it by $\frac{[nt]}{n}$. The moments of this sum are

$$E\left[B_t^{(n)}\right] = E\left[\frac{1}{\sqrt{n}}\sum_{i=1}^{[nt]}\xi_i(n)\right] = \frac{1}{\sqrt{n}}\sum_{i=1}^{[nt]}E\left[\xi_i(n)\right] = 0,$$

$$Var\left[B_t^{(n)}\right] = E\left[\left(\frac{1}{\sqrt{n}}\sum_{i=1}^{[nt]}\xi_i(n)\right)^2\right] = \frac{1}{n}\sum_{i=1}^{[nt]}E\left[(\xi_i(n))^2\right] = \frac{[nt]}{n} \to t,$$

so $B_t^{(n)}$ approximates standard Brownian motion.

The continuous time weighting kernel is adapted accordingly by averaging over the respective time interval. We hence get

$$z^{(n)}(t,s) := n\int_{s-\frac{1}{n}}^{s} z\left(\frac{[nt]}{n}, u\right) du. \tag{3.6}$$

Putting these parts together, Sottinen (2001) uses the following discrete approximation of fractional Brownian motion:

$$B_t^{H(n)} := \int_0^t z^{(n)}(t,s)dW_s^{(n)} = \sum_{i=1}^{[nt]} n\int_{\frac{i-1}{n}}^{\frac{i}{n}} z\left(\frac{[nt]}{n}, u\right) du \frac{1}{\sqrt{n}}\xi_i^{(n)}. \tag{3.7}$$

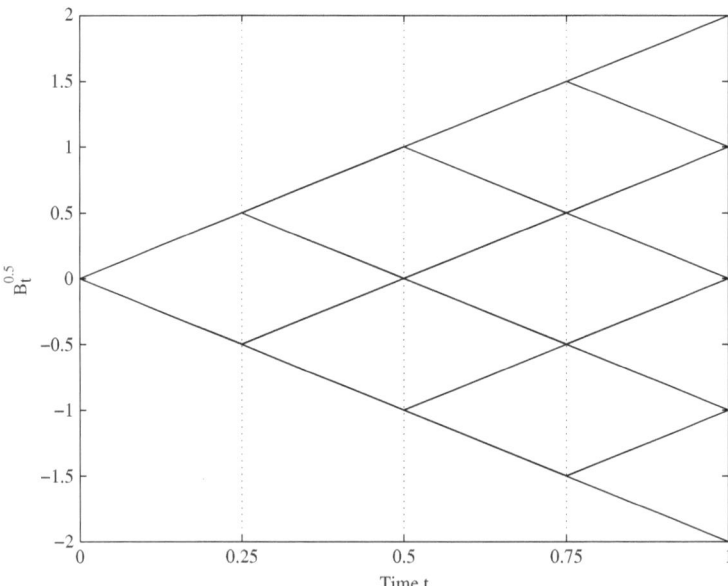

Fig. 3.1 Four-step approximation of classical Brownian motion ($H = 0.5$) for time $t = 1$

Note that for $H = \frac{1}{2}$, the weighting kernel $z(t, s)$ in (3.6) equals one and we get

$$B_t^{\frac{1}{2}(n)} := \int_0^t dW_s^{(n)} = \sum_{i=1}^{[nt]} n \int_{\frac{i-1}{n}}^{\frac{i}{n}} du \frac{1}{\sqrt{n}} \xi_i^{(n)} = \sum_{i=1}^{[nt]} \frac{1}{\sqrt{n}} \xi_i^{(n)} = B_t^{(n)}. \quad (3.8)$$

Hence we obtain a binomial random walk that is a mere sum of uncorrelated equally weighted random variables and which obviously approximates classical Brownian motion. Figure 3.1 depicts the according classical binomial tree, being symmetric and showing equidistant nodes at each step.

For $H \geq \frac{1}{2}$, Sottinen (2001) proves that the random walk of (3.7) indeed converges weakly to fractional Brownian motion. The weak convergence is proven by showing the identity of the moment properties as well as tightness and uses a general result of convergence of random variables that is due to Billingsley (1968). However, for the range of Hurst parameters standing for antipersistence, the original result by Billingsley (1968) proving tightness cannot be applied. Instead, a generalized version of the tightness criterion is needed, derived by (Genest et al. (1996), p. 332). We leave out the quite technical details and provide a not that rigorous, but all the more illustrative proof.

As known from probability theory, a random variable converges weakly to another random variable if their distribution functions converge pointwise at any point where the function is continuous (see e.g. Ash (1972)). Equivalently, a stochastic process—which is nothing else but a family of random variables indexed by time t—converges to another process, if for any time t the corresponding random variables converge. Hence, we have to look at the shape of the distribution functions of B_t^H on the one hand and $B_t^{H(n)}$ on the other hand.

Figure 3.2 depicts the distribution functions of a four-step as well as of a 18-step approximation in comparison to the exact distribution function of fractional Brownian motion at time $t = 1$. For a small number of steps the character of the approximating distribution function is of a step function kind, getting smoother the more steps are taken. Both for the antipersistent and for the persistent parameter choice, the convergence is easy to identify.

Having ensured convergence, we can now evolve binomial trees by the procedure mentioned above. The following figures depict the evolution of the binomial tress generated by the described random walk. Once more, we stress the fact, that for $H = \frac{1}{2}$, we would get the classical, recombining binomial tree. The situation for the cases where increments are correlated however is vitally different. Looking at Figs. 3.3 and 3.4, we observe that both the antipersistent tree and the persistent one are no longer recombining.

Additionally, though still being symmetric the outer branches are no longer

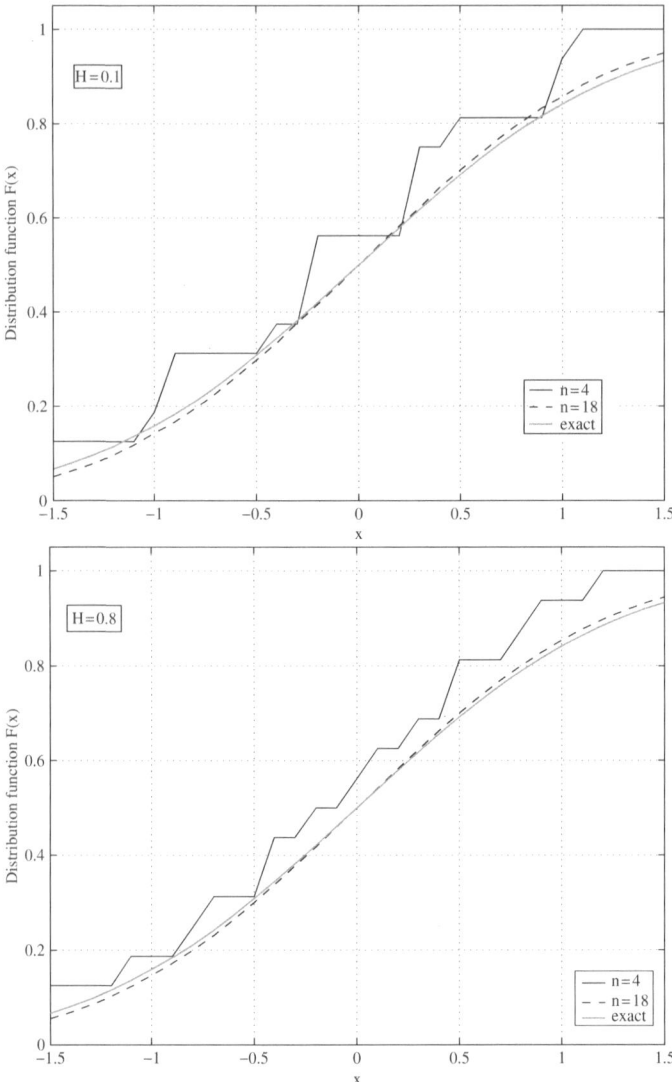

Fig. 3.2 Distribution functions of a four-step and a 18-step approximation of fractional Brownian motion for $H = 0.1$ (top) and $H = 0.8$ (bottom) in comparison to the exact distribution function.

steady lines but feature kinks. Thereby, Fig. 3.3 describes a concave envelope. This is due to the occurrence of antipersistence: The process always tends to invert the direction of the past that was recently struck in. Hence, it exhibits a kind of mean-reverting character. Additionally, the terminal nodes are no longer equidistant.

Meanwhile, for the persistent case, we look at Fig. 3.4 and observe similar

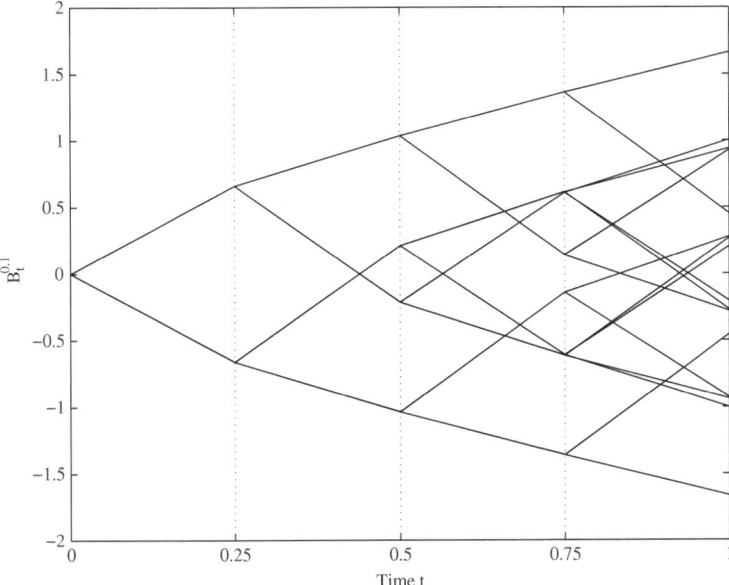

Fig. 3.3 Four-step approximation of fractional Brownian motion for time $t = 1$ and Hurst parameter $H = 0.1$

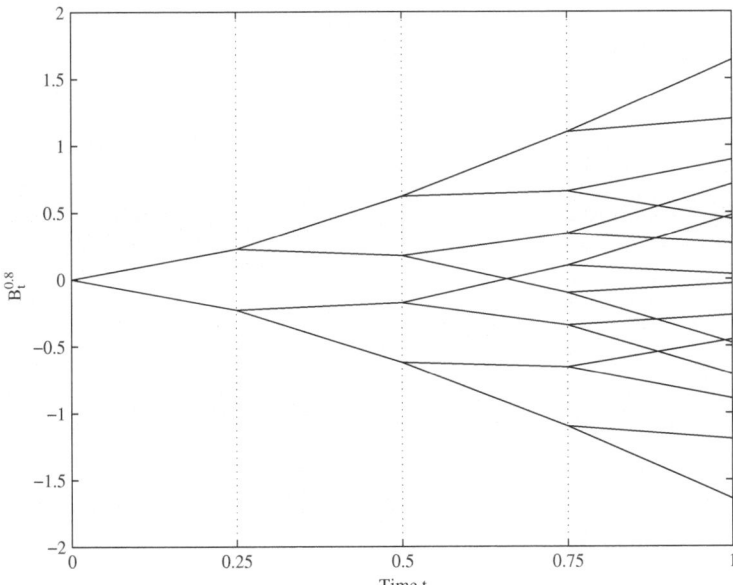

Fig. 3.4 Four-step approximation of fractional Brownian motion for time $t = 1$ and Hurst parameter $H = 0.8$

deviations from the standard case of the classical binomial tree. Again we perceive lack of recombination and equidistance, however, the envelope now shows a convex shape of nature. In this case, occurring persistence reinforces the chosen path and intensifies deviations from the mean. We will rediscover these phenomena when modeling the binomial price process in Sect. 3.3.

3.2 Binomial Approximation of the Conditional Moments of Fractional Brownian Motion

The approximation of the process of fractional Brownian motion allows us to proceed and also model conditional distributions of the process. In particular, we are interested in predicting the terminal value B_t^H by using the information of all steps up to a certain time t.

Recall the approximation procedure by Sottinen (2001) of B_T^H discussed in the preceding section, where we had

$$B_T^H = \sum_{i=1}^{[nT]} z^{(n)}\left(T, \frac{i}{n}\right) \frac{1}{\sqrt{n}} \xi_i^{(n)},$$

with $z^{(n)}(T, \frac{i}{n})$ as the weighting kernel of the ith step and with n as the number of approximation steps within one unit of time. Recall the finite-interval representation by Norros and Valkeila (1999) that underlies this approximation: It is evident that the conditional expectation of the process based on information up to time t can also be represented as a finite-interval of classical Brownian motion. We just cut off the independent future part. More formally, we obtain

$$
\begin{aligned}
E\left[B_T^H | \mathfrak{F}_t\right] &= E\left[\int_0^T z(T, s)\, dB_s | \mathfrak{F}_t\right] \\
&= \int_0^t z(T, s)\, dB_s + E\left[\int_t^T z(T, s)\, dB_s | \mathfrak{F}_t\right] \\
&= \int_0^t z(T, s)\, dB_s.
\end{aligned}
$$

Hence we can extend the idea of Sottinen (2001) by approximating the conditional expectation $\hat{B}_{T,t}^H$ by

$$\hat{B}_{T,t}^H = E\left[B_T^H | \mathfrak{F}_t\right] = \sum_{i=1}^{[nt]} z^{(n)}\left(T, \frac{i}{n}\right) \frac{1}{\sqrt{n}} \xi_i^{(n)},$$

that is, we use the same coefficients as for B_T^H but only summing up to time
t. Like for the fractional Brownian motion itself, we can also plot the values
of this conditional expectation along its possible paths. More precisely, after
each step, we calculate for each node the conditional mean of B_T^H given the
information up to this node. The outcome of this procedure is again some kind
of binomial tree which we call the conditional tree. The starting point of the
tree is the expectation of B_T^H conditional on time $t = 0$. As we do not have any
information at this point in time, this is of course equal to the unconditional
expectation of fractional Brownian motion, i.e. zero. On the other hand, at
all the terminal nodes of time $t = T$, the whole information is available and
$\hat{B}_{T,T}^H$ is just equal to B_T^H. So, the conditional tree is also a representation
of the evolution of fractional Brownian motion up to time T. However, the
nodes lying between zero and T now have a different meaning. While the
time t node of the unconditional tree converged to B_t^H, the respective node
of the conditional tree converges to $\hat{B}_{T,t}^H$. Each node indicates the mean of
all terminal nodes descending from it.

Taking a look at Figs. 3.5 and 3.6, one recognizes that, regarding the outer
shape of the binomial trees, the characteristics of the persistent and the antipersistent tree have interchanged. The conditional tree for Hurst parameters
$H < \frac{1}{2}$ now shows a convex or dispersing envelope, whereas the parameters
larger than one half standing for persistence yield a concave or contracting
shape of the conditional tree. Though this seems to contradict the explana-
tions given in the preceding section, this is actually not the case at all. To
see why, we have to interpret the meaning of one path within the conditional
tree which is nothing but the random evolution of the prediction for the terminal value. In the case of persistence, the first steps basically determine the
following ones. So, the leeway of the trajectory or the margin, respectively,
between to descending nodes is reduced after each additional step. Antipersistence however ascribes the main importance concerning predictability to
the last step. The latter fundamentally determines the terminal value. Increments with a lag larger than one step are less important. The first steps
consequently do not tell much about the terminal distribution. Contrarily,
the influence of the last draw on the terminal value is the crucial one. Consequently the margins increase along the tree.

Another distinctive feature between the conditional and the unconditional
tree is also worth mentioning: In the conditional tree, there is an easy link
between one node at an arbitrary step i and its immediate successors, as one
just has to add or subtract $z^{(n)}(T, \frac{i}{n})\frac{1}{\sqrt{n}}$, whereas the preceding summands
remain unchanged. Note that this property is not given within the uncondi-
tional tree, where all the historic ups and downs have to be rescaled by new
coefficients.

Given the tree of conditional means, it might prove interesting to investigate

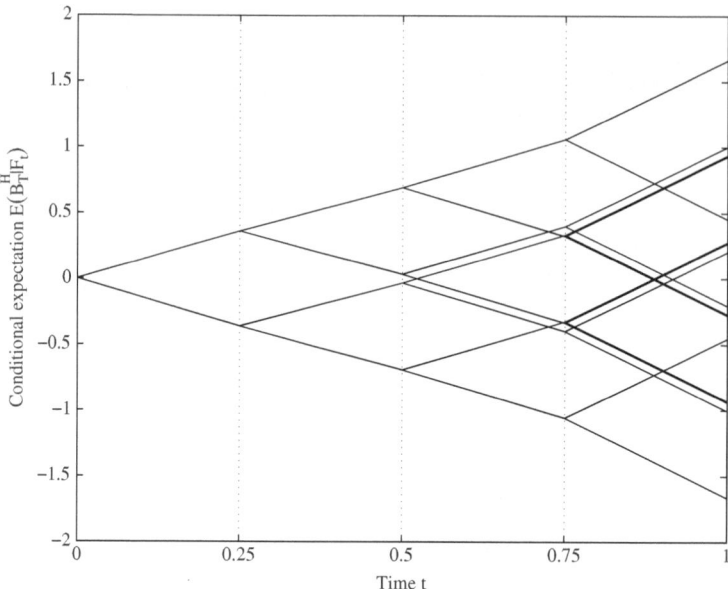

Fig. 3.5 Four-step approximation of the conditional mean of fractional Brownian motion for time $T = 1$ and Hurst parameter $H = 0.1$

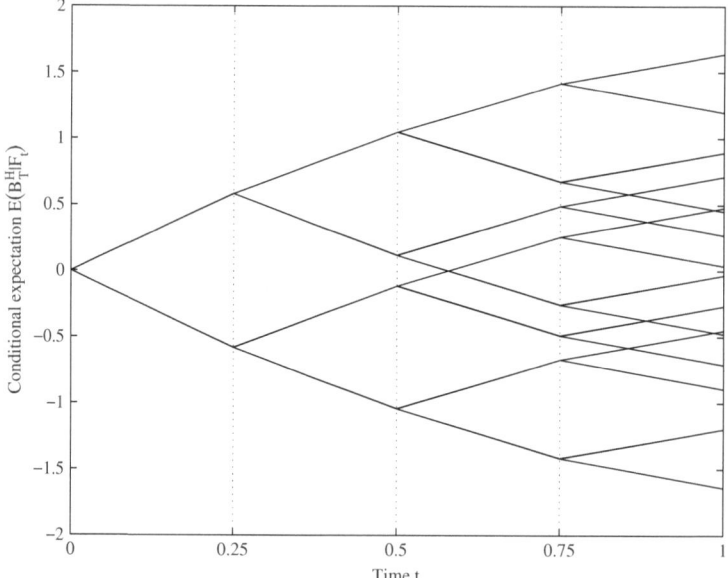

Fig. 3.6 Four-step approximation of the conditional mean of fractional Brownian motion for time $T = 1$ and Hurst parameter $H = 0.8$

the appropriate second moments. We hence look at the conditional variance defined by

$$\hat{\sigma}_{T,t}^2 = E\left[\left(B_T^H - \hat{B}_{T,t}^H\right)|\mathfrak{F}_t\right]^2.$$

Transforming this to our binomial model means measuring the deviations of the terminal nodes B_T^H around the time t node of the conditional tree. For a sufficiently large number of discretization steps, the results we obtain approximately equal the theoretical values of the according continuous time setting (for a derivation of the continuous time reference values, see Sect. 5.2). When comparing these results, one has to be aware of the differences that occur due to the fact that we cut off history. Meanwhile, one can show for the continuous time case that conditional variance within a certain prediction interval is nearly totally influenced by the historic interval of equal length (see Gripenberg and Norros (1996)). Consequently, we are able to ensure approximately good results by adjusting the information interval in the required manner.

Number of steps / Hurst parameter	$n = 4$	$n = 8$	$n = 12$	$n = 16$	$n = \infty$
$t = \frac{1}{4}T$					
$H = 0.1$	0.6095	0.6352	0.6448	0.6513	0.6866
$H = 0.3$	0.8525	0.8432	0.8375	0.8336	0.7837
$H = 0.5$	0.2500	0.2500	0.2500	0.2500	0.2500
$H = 0.7$	0.5462	0.5726	0.5819	0.5868	0.6009
$H = 0.9$	0.2241	0.2473	0.2550	0.2589	0.2701
$t = \frac{1}{2}T$					
$H = 0.1$	0.5007	0.5213	0.5328	0.5415	0.6014
$H = 0.3$	0.6568	0.6468	0.6409	0.6369	0.6012
$H = 0.5$	0.5000	0.5000	0.5000	0.5000	0.5000
$H = 0.7$	0.2906	0.3098	0.3167	0.3203	0.3316
$H = 0.9$	0.0907	0.1038	0.1082	0.1103	0.1167
$t = \frac{3}{4}T$					
$H = 0.1$	0.3659	0.3930	0.4053	0.4129	0.5066
$H = 0.3$	0.4327	0.4269	0.4225	0.4193	0.3910
$H = 0.5$	0.2500	0.2500	0.2500	0.2500	0.2500
$H = 0.7$	0.0968	0.1088	0.1133	0.1157	0.1236
$H = 0.9$	0.0181	0.0248	0.0271	0.0282	0.0315

Table 3.1 Conditional variance $\hat{\sigma}_{t,1}^{2(n)}$ of fractional Brownian motion for different points in time t and different Hurst parameters H

Table 3.1 depicts the values of the conditional variances based on a different number of approximating steps. The comparison with the continuous time limit case (we refer the reader to Chap. 5 to see how the latter can be calculated) shows that the speeds of convergence differ eminently. The speed

depends both on the Hurst parameter and on the ratio between the lengths
of the observation and the prediction interval, respectively. Moreover, one
becomes aware of the fact that for a fixed number of steps there is a tradeoff
when choosing the ratio between the observation and the prediction window.
A more detailed expansion of the future distribution has to be paid by less
information about the past. For the parameters standing for persistence or
weak antipersistence, it seems to be favorable to have an information period
being at least as long as that of prediction. In the case of strong antipersistence however—where as we recall only the most recent historic events
influence the future distribution—the main importance should be attached
to an exact mapping of the future distribution of the process.

Two more important mathematical features should be mentioned at this
point. On the one hand, the fact that the conditional and the unconditional
tree differ at all, has a crucial meaning: Evidently, the process of fractional
Brownian motion is no longer a martingale, otherwise the future prediction in
time t should equal the present value and both trees would coincide. On the
other hand, the prediction not only depends on the last, but on all historic
random realizations, so the process also is no longer Markovian. In Sect. 3.4,
we will investigate, how these properties affect the usage of fractional Brownian motion in financial models.

3.3 Binomial Approximation of a Geometric Fractional Price Process

We have introduced fractional Brownian motion as the source of randomness
in the discrete framework and got to know about some characteristic features.
We now keep this discrete time vantage point and have a look at geometric
fractional Brownian motion, which is the stochastic process S_t satisfying the
following differential equation:

$$dS(t) = \mu S(t)dt + \sigma S(t)dB_t^H$$

where dB_t^H is the increment of fractional Brownian motion.
From the preceding sections we already know that this differential equation does not yet determine all properties of the process as we can trans-
form this equation into an integral equation and then have to interpret the
stochastic fractional integral—either in the ordinary pathwise sense or in the
Wick sense.
We recover an analogy of this aspect in the discrete modelling of geometric
fractional Brownian motion. As in the well-known case of classical geometric
Brownian motion we generate a recursive multiplicative tree, starting with a
value S_0 and introduce the recursion law

$$S_n = S_{n-1} \bullet (1 + \mu_n + X_n),$$

where $\bullet$ is either the ordinary product or a discretization of the Wick product and X_n is an approximation of the fractional Brownian increment as in the section before. For the reasons mentioned there, we restrict ourselves to the approximation procedure by Sottinen (2001) when modeling X_n.

Let us first present the idea of the discrete Wick product. Following Holden et al. (1996), any square-integrable random variable X that is defined on a n-fold Cartesian product of $\{-1, 1\}$ has a unique representation using independently and identically distributed binary random variables ξ_i

$$X = \sum_{A \subset \{1,\dots,n\}} \left(X(A) \prod_{i \in A} \xi_i \right); \quad X(A) \in \mathbb{R}.$$

This is called the Walsh decomposition. Using this result, it is sufficient to define the discrete Wick product of two products of an arbitrary subset of $\xi_i, i \in \{1. \dots, n\}$ and extending it to more general random variables using their Walsh decomposition. According to this, we define the discrete Wick product—denoted by $\diamond_d$—as follows (see Bender (2003a)):

$$\prod_{i \in A} \xi_i \diamond_d \prod_{j \in B} \xi_j = \begin{cases} \prod_{i \in A \cup B} \xi_i, & \text{if } A \cap B = \emptyset \\ 0, & \text{otherwise.} \end{cases}$$

Verbalizing this definition we can say that the discrete Wick product vanishes if the two factors have at least one generating random variable in common. If none of the generators coincide, the wick product equals the ordinary product. This property becomes crucial when investigating the construction of discrete geometric fractional Brownian motion. We exemplify this with a short calculation.

Recall the representation of Sottinen discussed above: Fractional Brownian motion was approximated by the sum

$$B_T^H = \sum_{i=1}^{[nT]} z^{(n)} \left(T, \frac{i}{n} \right) \frac{1}{\sqrt{n}} \xi_i^{(n)} = \sum_{i=1}^{[nT]} k^{(n)}(T, i) \xi_i^{(n)},$$

where $k^{(n)}(T, i) = z^{(n)}(T, \frac{i}{n}) \frac{1}{\sqrt{n}}$. We regard only two steps of recursion and use the step size 1, that is $t_1 = 1$, $t_2 = 2$. Furthermore, we abstain from a drift component μ. For sake of simplicity, we omit the superscript (n). Corresponding to Chap. 2, we have the following representation of the fractional Brownian motion:

$$B_0^H = 0, \qquad B_1^H = k(1,1)\xi_1, \qquad B_2^H = k(2,1)\xi_1 + k(2,2)\xi_2.$$

Therefore, the Brownian increments $dB_j = B_j - B_{j-1}$ are

$$dB_1^H = k(1,1)\xi_1, \qquad dB_2^H = (k(2,1) - k(1,1))\xi_1 + k(2,2)\xi_2.$$

With $S_0 = 1$, the price process of the geometric fractional Brownian motion in the pathwise sense, denoted by $S_t^{(P)}$, develops as follows:

$$
\begin{aligned}
S_0^{(P)} &= 0 \\
S_1^{(P)} &= S_0^{(P)}(1 + dB_1^H) = 1 + k(1,1)\xi_1, \\
S_2^{(P)} &= S_1^{(P)}(1 + dB_2^H) \\
&= (1 + k(1,1)\xi_1)\,(1 + (k(2,1) - k(1,1))\xi_1 + k(2,2)\xi_2) \\
&= 1 + k(2,1)\xi_1 + k(2,2)\xi_2 + k(1,1)k(2,2)\xi_1\xi_2 \\
&\quad + k(1,1)(k(2,1) - k(1,1))\xi_1^2.
\end{aligned}
$$

Note that by construction, ξ_1^2 is deterministic with value 1, so that the term $k(1,1)(k(2,1) - k(1,1))$ contributes to the drift of the process. This is the analogy to the continuous time setting where the fractional integral in the pathwise sense yields a non-zero expected value (see Sect. 2.4). In contrast, when we look at the evolution of the geometric fractional Brownian motion in the Wick sense, denoted by $S_t^{(W)}$, we obtain

$$
\begin{aligned}
S_0^{(W)} &= 0, \\
S_1^{(W)} &= S_0^{(W)} \diamond_d (1 + dB_1^H) = 1 + k(1,1)\xi_1, \\
S_2^{(W)} &= S_1^{(W)} \diamond_d (1 + dB_2^H) \\
&= (1 + k(1,1)\xi_1) \diamond_d (1 + (k(2,1) - k(1,1))\xi_1 + k(2,2)\xi_2) \\
&= 1 + k(2,1)\xi_1 + k(2,2)\xi_2 + k(1,1)k(2,2)\xi_1\xi_2.
\end{aligned}
$$

In this case, the Wick product eliminates the squared term by its definition and the drift remains unchanged. We stress again the parallel to the continuous framework where we received an expected value of zero for the fractional integral in the Wick sense. Actually, this was the motivation for introducing the stochastic integration calculus based on Wick products. Obviously, the discretization of the concept reveals the same properties.

It is important to check whether this alternative way of multiplication fundamentally changes the basic properties of the process, i.e. the type of its distribution. Comparing the corresponding binomial trees in Figs. 3.7–3.9, plotted for four time steps without a drift component, we see that this is not the case. Moreover, for any of the three qualitatively different cases of antipersistence ($H < \frac{1}{2}$), independence ($H = \frac{1}{2}$) and persistence ($H > \frac{1}{2}$), the Wick binomial tree resembles that of ordinary multiplication. Thereby, we observe features that correspond both to the preceding shapes of fractional

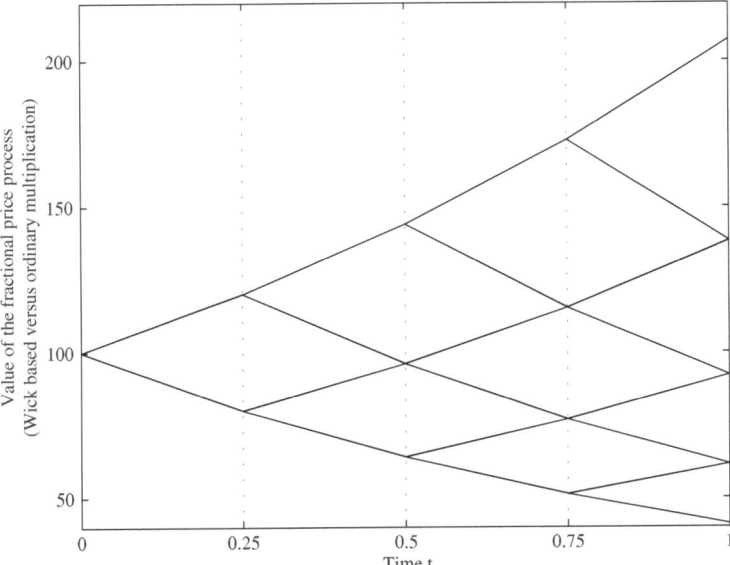

Fig. 3.7 Binomial approximation of the geometric Brownian motion based on the Wick product and on the ordinary product

Brownian motion and to the distributional properties of the respective continuous time processes. Again, antipersistence comes along with a concave evolution of the variance over time whereas persistence generates a convex border of the binomial tree. As in the section before, we obtain a recombining tree only for the case $H = \frac{1}{2}$. Having stated approximative normality for the fractional Brownian motion, one can observe—at least for the persistent and for the independent case—the log-normality of geometric fractional Brownian motion.

Nevertheless, even for such a small number of steps, distinctions appear. The latter are clarified in Figs. 3.8 and 3.9 by the different shapes of the trees. The dashed lines represent the Wick-product-based trees, whereas the solid lines depict the trees using ordinary multiplication. Obviously, the Wick product effectuates a correction of the values generated by pathwise multiplication, yielding larger values for the antipersistent case and smaller values for parameters $H > \frac{1}{2}$. Against the background of the preceding chapter, these differences are not at all surprising, but the exact analogon to the interrelation between the two types of fractional integrals stated in Table 2.1.

Meanwhile, for the independent case, the binomial trees do not differ at all. This is due to the fact that the Brownian increments $dB_n^{\frac{1}{2}}$ are represented only by ξ_n and do not depend on the preceding $\xi_{j,\,\{j<n\}}$. Therefore, no squared terms occur in the recursion formula, so the Wick product and the ordinary product yield the same results. This is again consistent with the according

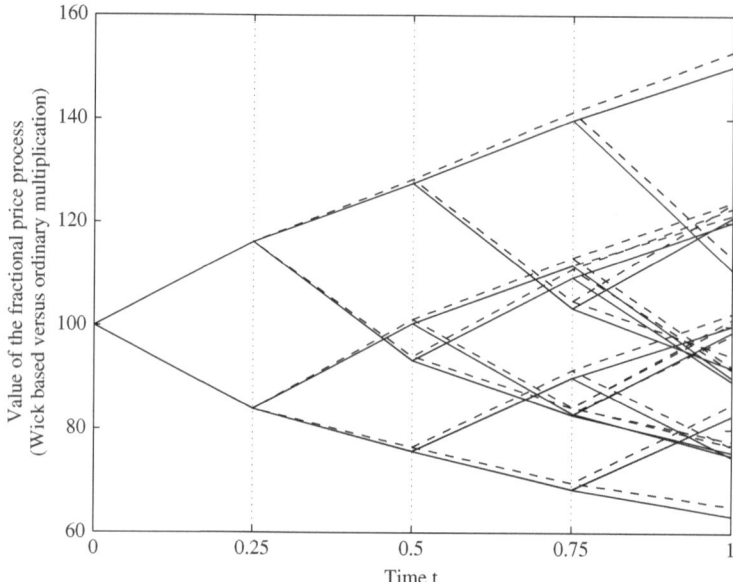

Fig. 3.8 Binomial approximation of the geometric fractional Brownian motion for the antipersistent case of $H = 0.1$ based on the Wick product (dashed line) and on the ordinary product (solid line)

proposition of the continuous time setting, where the fractional integral of Wick type and the pathwise integral coincide for the case of serial independence, that is for $H = \frac{1}{2}$.

The preceding considerations helped to impart intuition to what the process of fractional Brownian motion looks like and by what key features it is constituted. However, the most important application of the well-known binomial trees that model classical geometric Brownian motion, is, that by absence of arbitrage one can also model prices of derivative assets. In the following chapter, we will see that for Hurst parameters $H \neq \frac{1}{2}$, this possibility is no longer given in the fractional framework, at least as long as one does not impose further restrictions. Though this seems to be disillusioning concerning a further use of fractional binomial trees, our modified discrete setting will prove valuable when we introduce a preference based pricing approach. In particular, the fact, that we can easily model conditional moments, will turn out to be helpful.

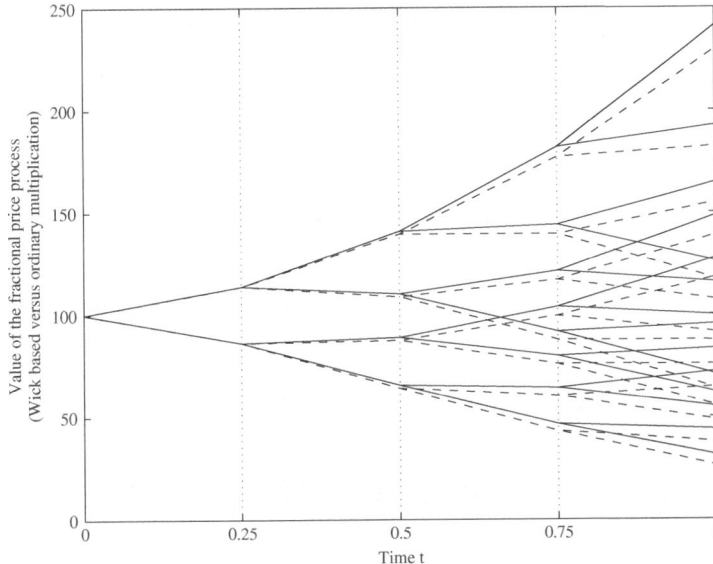

Fig. 3.9 Binomial approximation of the geometric fractional Brownian motion for the persistent case of $H = 0.8$ based on the Wick product (dashed line) and on the ordinary product (solid line)

3.4 Arbitrage in the Fractional Binomial Market Setting and Its Exclusion

Within the preceding sections, we learned that if the fractional price process is expanded using ordinary multiplication, the mean of the price process is not equal to its starting value. this was true even if no deterministic trend was given. Obviously, this seems to be at odds with our idea of a fair game, as a systemic upward bias occurs and one should expect arbitrage possibilities. Hence, the discrete Wick product was introduced and it was shown that the resulting process accounts for that phenomenon and yields an unbiased mean of the process. Yet, if one is interested in conditional statements, it turned out that conditional expectation based on all available historic information does not equal the present value of the process, or to put it differently, the martingale property is not given. So, despite the improvement of an unconditionally unbiased process, the game becomes unfair as soon as information about the past can be exploited. Consequently, one should also expect arbitrage within the Wick-product-based setup. Indeed, taking for example a look at Fig. 3.9, one recognizes both for the ordinary product and the Wick-product-based process one conspicuous peculiarity: Concerning the nodes before last, one can see that—starting from the top node—both descending branches show upward direction. That means, in absence of a riskless interest rate, a one-step buy and hold strategy will always pay and promises a riskless gain.

In the following, we will present explicit proofs for the existence of arbitrage possibilities within the unrestricted setting. The results were provided by Sottinen (2001) for ordinary multiplication and by Bender (2003a) for Wick multiplication.

The binary model for geometric fractional Brownian motion was presented in the preceding section as a multiplicative process using the ordinary product or the Wick product, respectively. Again, the assets only change their value at discrete points in time $0 = t_0 < t_1 < \ldots < t_n = T$. The framework is extended to a binary market model by additionally introducing a riskless asset, or bond $A_j^{(n)}$, where the subscript j denotes the value at time t_j and the superscript n as in the sections before indicates the fineness of the approximation. Combined, we obtain the following setting consisting of the riskless asset with dynamics

$$A_j^{(n)} = (1 + r^{(n)}) A_{j-1}^{(n)},$$

as well as the risky stock with dynamics

$$S_j^{(n)} = \left(1 + \mu^{(n)} + X_j^{(n)}\right) \circ S_{j-1}^{(n)},$$

where $X_j^{(n)}$ is the corresponding n-th approximation of the increment of fractional Brownian motion, that is $dB_{t_j}^H$ and $\circ$ stands for the particular way, in which multiplication is carried out. From the sections above, we derive that this random variable $X_j^{(n)}$ is binary and has the following representation:

$$X_j^{(n)} = \sigma \left(B_{t_j}^{H(n)} - B_{t_{j-1}}^{H(n)} \right) \tag{3.9}$$

$$= \sigma \sqrt{n} \left(\sum_{i=1}^{j} \int_{\frac{i-1}{n}}^{\frac{i}{n}} z\left(\frac{j}{n}, s\right) ds\, \xi_i^{(n)} - \sum_{m=1}^{j-1} \int_{\frac{m-1}{n}}^{\frac{m}{n}} z\left(\frac{j-1}{n}, s\right) ds\, \xi_m^{(n)} \right).$$

Sottinen (2001) introduces the shorthand notations

$$k(j,i) := k^{(n)}(j,i) = \sqrt{n} \int_{\frac{i-1}{n}}^{\frac{i}{n}} z\left(\frac{j}{n}, s\right) ds$$

as well as

$$f_{j-1}(\xi_1, \ldots, \xi_{j-1}) := \sum_{i=1}^{j-1} (k(j,i) - k(j-1,i)) \xi_i$$

and rewrites (3.9) by

$$X_j^{(n)} = \sigma \left(k(j,j) \xi_j + f_{j-1}(\xi_1, \ldots, \xi_{j-1}) \right).$$

For each step j, based on the knowledge of the complete past, there are only two possible values of the binary random variable $X_j^{(n)}$ which are denoted by $d_j^{(n)}$ (for $\xi_j^{(n)} = -1$) and $u_j^{(n)}$ (for $\xi_j^{(n)} = +1$).

We choose the interest rate $r^{(n)}$ and the drift rate $\mu^{(n)}$ to be the n-th part of the according constants of the respective continuous time setting. Then, the discrete model converges to the continuous framework as n tends to infinity (see Sottinen (2001)). In the following, we will omit the superscript n, wherever possible.

Concerning the absence of arbitrage possibilities, it is necessary to ensure, that for any time step we have

$$S_j^n(\xi_j = -1) - S_{j-1}^n < rS_{j-1}^n < S_j^n(\xi_j = +1) - S_{j-1}^n, \tag{3.10}$$

that is, the return of the risky asset exceeds the riskless interest rate in the case $X_j^n = u_j^{(n)}$ and falls below the riskless rate in the case $X_j^n = d_j^{(n)}$.

We first address ourselves to the case where ordinary multiplication is used. In this case, the increment $S_j^n - S_{j-1}^n$ can be rewritten:

$$S_j^n - S_{j-1}^n = \mu S_{j-1}^n + X_j^n S_{j-1}^n.$$

Using this, (3.10) can be simplified and one gets the relation

$$d_j < r - \mu < u_j. \tag{3.11}$$

Evidently, the existence of arbitrage can be proven by one example contradicting relation (3.11). For this purpose, (Sottinen (2001), p. 353) picks the two special cases where the historical path either always moved upwards or downwards, that is, he investigates the sequences $(\xi_1, \xi_2, \ldots, \xi_{j-1}) = (\pm 1, \pm 1, \ldots, \pm 1)$. The according inequalities (3.11) then can be rewritten as follows:

$$\sigma\left(-k(j,j) + f_{j-1}(\pm 1, \ldots, \pm 1)\right) < r - \mu < \sigma\left(k(j,j) + f_{j-1}(\pm 1, \ldots, \pm 1)\right).$$

In order to prove arbitrage, it is necessary and sufficient that at least one of these inequalities fails. Exploiting the symmetry of the two equations, the problem can be reduced to the proof of

$$f_{j-1}(1, \ldots, 1) - k(j,j) \geq 0. \tag{3.12}$$

The latter inequality is verified to hold for all j larger than a critical step number N_H, only depending on the Hurst parameter and tending to infinity as H approaches one half. The result can be derived by a series of quite technical arguments and is shown in (Sottinen (2001), p. 353 et seq.). The author also provides an explicit arbitrage possibility: Suppose the difference $r - \mu$ to be negative and let the historic path up to step $j > N_H$ be strictly upwards moving, that is $(\xi_1, \xi_2, \ldots, \xi_{j-1}) = (+1, +1, \ldots, +1)$. Hence we obtain

$$d_j = f_{j-1}(1, \ldots, 1) - k(j, j),$$
$$u_j = f_{j-1}(1, \ldots, 1) + k(j, j),$$

which are—as relation (3.12) holds—both nonnegative, and we obtain

$$r - \mu < 0 \leq d_j < u_j, \tag{3.13}$$

so the fundamental no arbitrage relation (3.11) is violated. In this situation, an investor will buy one stock at step $j - 1$ at the price S_{j-1} and borrow the same amount paying the riskless interest rate r. In the worst case, ξ_j equals minus one and the stock moves what we call downwards and takes the value $S_j = S_{j-1}(1 + \mu + d_j)$. But, taking a look at (3.13), $\mu + d_j$ already exceeds the riskless interest rate, so the riskless gain is guaranteed. If, on the other hand, the difference $r - \mu$ is positive, one can—with positive probability—exploit the path $(\xi_1, \xi_2, \ldots, \xi_{j-1}) = (-1, -1, \ldots, -1)$, by short-selling the stock and investing in the riskless asset.

In the case of the setting based on the Wick product, the proof of an arbitrage possibility is a little bit more complicated. Bender (2003a) shows that the increment $S_j^n - S_{j-1}^n$ now satisfies:

$$S_j^n - S_{j-1}^n = S_0 X_j^n + O\left(n^{-(1 \wedge 2H)}\right). \tag{3.14}$$

In order to account for the term of an order that depends on N, the coefficients $k(j, i)$ have to hold a stronger condition than (3.12) if arbitrage should be possible. With

$$x^n(j, i) = \begin{cases} k^n(j, i) - k^n(j - 1, i), & \text{if } i < j \\ k^n(j, j) & , & \text{if } i = j. \end{cases}$$

This condition can be formulated as follows: If there are lower bounds $y_l(j, i)$ and upper bounds $y_u(j, i)$, both independent of n, so that

$$n^{-H} y_l(j, i) \leq |x^n(j, i)| \leq n^{-H} y_u(j, i) \tag{3.15}$$

$$\text{and} \qquad \sum_{i=1}^{j-1} y_l(j, i) > y_u(j, j), \tag{3.16}$$

then there exists an arbitrage possibility in the market. Bender (2003a) explicitly derives these upper and lower bounds to be

$$y_l(j,i) = \begin{cases} C_H\sigma \left[(j+1+i)^{H-\frac{1}{2}} - (j-i)^{H-\frac{1}{2}} \right], & \text{if } i < j \\ \frac{C_H\sigma}{H+\frac{1}{2}}, & \text{if } i = j. \end{cases}$$

and

$$y_u(j,i) = \begin{cases} \frac{C_H\sigma}{\frac{3}{2}-H} \left(\frac{j}{i-1} \right)^{H-\frac{1}{2}} \left[(j+1+i)^{H+\frac{1}{2}} - (j-i)^{H+\frac{1}{2}} \right], & \text{if } i > 1 \\ \frac{C_H\sigma}{\frac{3}{2}-H} j^{2H-1}, & \text{if } i = 1. \end{cases}$$

We now investigate the same state of nature as in the example by Sottinen (2001) mentioned above. Suppose all preceding random draws ξ_i, $i < j$ to be positive. The question is, whether a realization $\xi_j = -1$ is able to make the value of the stock move downwards or not. Inserting $\xi_i = +1$, $i < n$ into (3.14), one obtains

$$S_j^n - S_{j-1}^n = S_0 \sum_{i=1}^{j-1} |x^n(j,i)| + S_0 x^n(j,j)\xi_j + O\left(n^{-(1\wedge 2H)}\right)$$

With the bounds defined above satisfying (3.15) and (3.16), this can be further specified (see Bender (2003a), p. 140):

$$S_j^n - S_{j-1}^n \geq n^{-H} \left[S_0 \sum_{i=1}^{j-1} y_l(j,i) - S_0 y_u(j,j) O\left(n^{-[(1-H)\wedge H]}\right) \right]$$
$$> 0,$$

for a sufficiently high level of fineness n. Consequently, even a negative evolution in time step j (that is $\xi_j = -1$) cannot invert the upward trend. Hence, a one-step buy-and-hold strategy will always yield a positive gain. The described situation occurs with positive probability, as we are given a finite number of states (paths). So, we have an arbitrage possibility also in the binomial Wick-based setting.

Albeit the existence of arbitrage possibilities within the framework described above seems disillusioning at first glance, the situation is not as bad as it seems. The given examples all postulated that the investor can react as fast as the market. This means, one can make transactions at each node of the random path and thereby realize a one-step buy-and-hold strategy. Meanwhile, it might not seem to be too restrictive if one introduces a minimal delay between two consecutive transactions of one and the same investor. Given a multitude of investors, we assert that within this minimal processing time, other investors do make transactions.
We forestall a result of the continuous time case of Chap. 4 where an analogous problem will occur. There, we will restrict the in the way that the market still evolves continuously, whereas each single investor can only trade

at discrete points in time. This modified framework where investors cannot be as fast as the market is free of arbitrage. It is due to Cheridito (2003) and will be recalled later on.

It is this basic idea that we also apply to our discrete setting. Putting it into the framework of binomial trees, a single investor cannot catch two consecutive nodes: Even if he sells immediately after having bought an asset, he would have missed a number of transaction nodes caused by the multitude of investors. It can be shown that this assumption is already sufficient to rule out arbitrage possibilities. More formally: Assume we have an arbitrage possibility within a certain interval, as in the examples before, driven by an extreme realization of the historic path. Then one can always introduce a number of intra-interval nodes caused by the rest of the market, so that the arbitrage opportunity vanishes. We do not give an explicit proof but provide an illustrative example instead.

Figure 3.10 depicts a market where a single investor can only make transactions at each time unit but not on the nodes in-between. The above picture where no additional steps of other market participants lie in-between allows for a one-step buy-and-hold arbitrage, buying the stock in time $t = 3$ and selling it in $t = 4$. In the lower picture, an additional node of market transactions is introduced that cannot be exploited by an investor buying the stock in time $t = 3$. Consequently, the same strategy can yield a loss in $t = 4$ and therefore no longer exhibits an arbitrage possibility. Note that the number of additional nodes between two consecutive transactions of one investor being necessary to exclude arbitrage depends on both the Hurst parameter and on the amount of historic information that is available. Generally speaking, both a higher level of persistence and more information about the past increase the number of necessary steps.

In order to avoid misunderstandings, we stress once more the basic idea of how arbitrage can be excluded. The binomial framework per se is already a discrete framework: that is, transactions cannot be proceeded infinitely fast. However, within the unrestricted setting each investor can at least be as fast as the market: that is one can catch any change in value of the traded assets. The additional restriction we impose, limits the relative speed of consecutive transactions. Due to the multitude of market participants, investors cannot be as fast as the market. So, existing arbitrage possibilities cannot be exploited, or in other words, the respective strategies are not admissible.

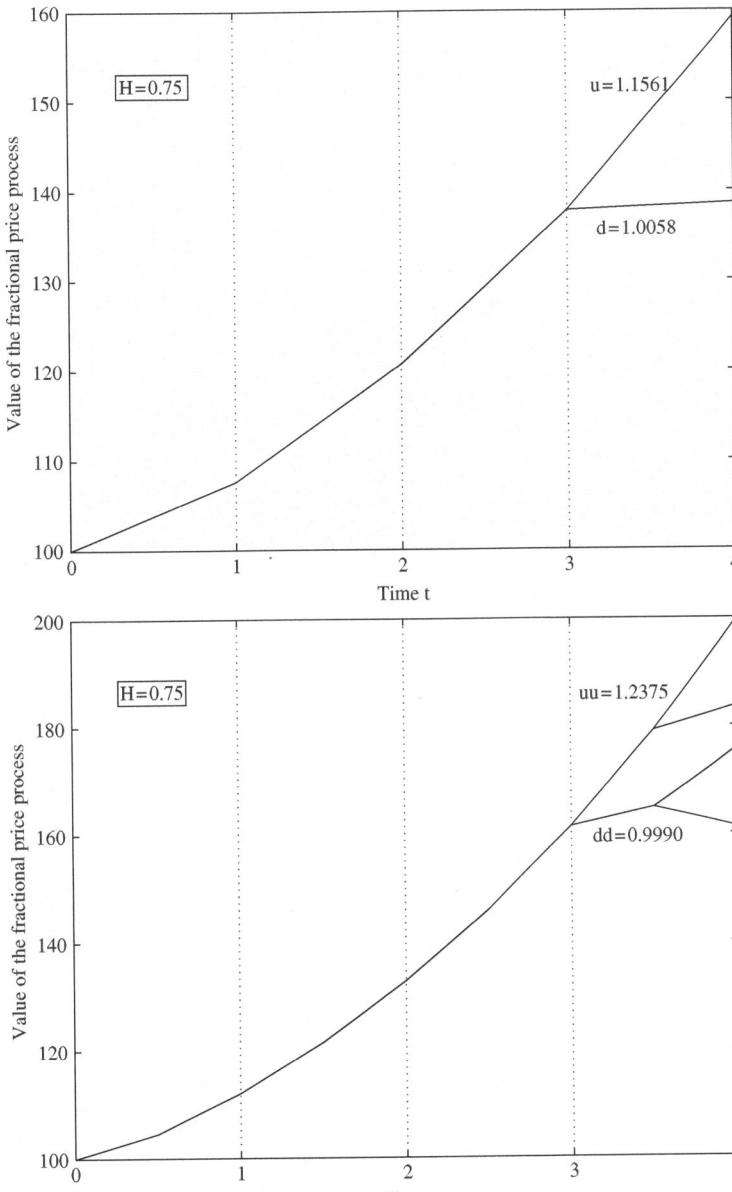

Fig. 3.10 Exclusion of arbitrage by restricting trading strategies. (Investor can only make transactions at nodes on a dashed line.)

Chapter 4
Characteristics of the Fractional Brownian Market: Arbitrage and Its Exclusion

We have seen that fractional Brownian motion is a generalization of classical Brownian motion and hence it is naturally closely related to the latter. Moreover, both models of randomness have important properties in common: first of all their Gaussian character. However, the main difference is a crucial one: For all Hurst parameters $H \neq \frac{1}{2}$, fractional Brownian motion exhibits serial correlation implicating sort of a memory of the process. This information about the past makes it in turn possible to say something about the future, that is, predictability comes into play. We will see that this predictability can imply arbitrage possibilities when we use fractional Brownian motion as source of randomness in our market setup. However, one can impose restrictions to trading strategies by which arbitrage can be excluded.

For the discrete time binomial approach we could already observe the occurrence of arbitrage possibilities if trading strategies were not sufficiently restricted (see Chap. 3). But also for the continuous time case, models using fractional Brownian motion revealed some problems. It was Rogers (1997) who made the case that fractional Brownian motion was an unsuitable candidate for usage in financial models: for all Hurst parameters $H \neq \frac{1}{2}$ he had derived existence of arbitrage possibilities in a fractional Bachelier type model. However, his setting being linear and passing on the existence of a drift, the question of generality concerning his results should need some further investigation.

We now introduce the continuous time market setup which we want to focus on. Based on the definition of fractional Brownian motion, we look at a fractional Brownian market consisting of a riskless asset or bond A_t with dynamics

$$dA_t = r A_t \, dt, \qquad (4.1)$$

as well as of a risky asset or stock S_t with dynamics

$$dS_t = \mu S_t \, dt + \sigma S_t \, dB_t^H. \qquad (4.2)$$

S. Rostek, *Option Pricing in Fractional Brownian Markets.*
Lecture Notes in Economics and Mathematical Systems.
© Springer-Verlag Berlin Heidelberg 2009

The process satisfying the latter equation is called geometric fractional Brownian motion. Unless otherwise stated, the parameters r, μ and σ are assumed to be constant, symbolizing the riskless interest rate as well as the drift and the volatility of the stock. The mathematical interpretation of (4.2) depends on the assumed integration theory, by name pathwise integration or Wick-based integration, respectively.

In this chapter we investigate the characteristics of the fractional Brownian market when it is based on the continuous stochastic process. In particular, we discuss in Sect. 4.1 thoroughly the problem of arbitrage. Furthermore we will present in Sect. 4.2 the diverse approaches that have been stressed during the last few years to overcome the existence of arbitrage. We then focus on the problem of dynamical completion showing regard to the specific character of fractional Brownian motion. As a result, Sect. 4.4 will show that the renouncement of continuous tradability combined with the transition to a preference-based pricing poses a passable way to give a sensible meaning to the problem of asset pricing in fractional Brownian markets.

4.1 Arbitrage in the Unrestricted Continuous Time Setting

Following the debate of the history, we first address ourselves to a market model based on pathwise integration. In the second part of this section we proceed and consider the more challenging Wick product integration concept. It will turn out that both concepts admit arbitrage as long as one does not delimit the type of trading strategies an investor can realize.

4.1.1 Arbitrage in the Continuous Setting Using Pathwise Integration

We know from the preceding chapters that fractional Brownian motion and hence all processes driven by it are not semimartingales. Even worse, if (4.2) is interpreted in the pathwise/Stratonovich sense, the corresponding stochastic integral does not have zero expectation which already suggests the possibility of riskless gains. These obvious shortcomings of a fractional Brownian market model based on pathwise integration were primarily worked out by Shiryayev (1998). Further examples of arbitrage are given by Dasgupta and Kallianpur (2000) or Bender (2003a).

For reasons of simplicity, Shiryayev (1998) discusses a financial model where the drift of the risky asset equals the interest rate of the riskless asset and the volatility equals one. That is, we have a bond following

$$dA(t) = rA(t)dt, \tag{4.3}$$

and a stock following

$$dS_t = rS_t\, dt + S_t\, \delta B_t^H. \tag{4.4}$$

The definition of the stochastic integral the author uses is a bit different to other definitions of pathwise integration (see Chap. 2). Substantially, it is also a Riemann type limit using ordinary multiplication. The integration concept can be embedded into the more general integration theory of fractional Stratonovich integration (see Bender (2003a), p. 72–76). It is worth looking at the crucial property of the stochastic integration calculus Shiryayev makes use of: It is the chain rule (see Shiryayev (1998), p. 3) given by

$$F(T, B_T^H) = F(t, B_t^H) + \int_t^T \frac{\partial F}{\partial s}\, ds + \int_t^T \frac{\partial F}{\partial B_s^H}\, \delta B_s^H,$$

or, respectively, formulated as differential equation

$$dF(s, B_s^H) = \frac{\partial F}{\partial s}\, ds + \frac{\partial F}{\partial B_s^H}\, \delta B_s^H.$$

From this, it follows immediately that the explicit equations of the basic market assets have to be of the form

$$A_t = A_0 e^{rt},$$

and

$$S_t = S_0 e^{rt + B_t^H}.$$

From now one, the initial values A_0 and S_0 of the two basic assets are assumed to be one. Let X_t^π be the value of the portfolio based on the strategy $\pi_t = (\beta_t, \gamma_t)$, where β_t and γ_t denote the current positions held in the bond or the stock, respectively. Hence, this portfolio value is given by

$$X_t^\pi = \beta_t A_t + \gamma_t S_t, \tag{4.5}$$

and the strategy is called self-financing, if

$$dX_t^\pi = \beta_t\, dA_t + \gamma_t\, dS_t.$$

Now, the following dynamic strategy is considered:

$$\beta_t = 1 - e^{2B_t^H},$$
$$\gamma_t = 2\left(e^{B_t^H} - 1\right).$$

Inserting this into (4.5), one obtains

$$X_t^\pi = \left(1 - e^{2B_t^H}\right) e^{rt} + 2 \left(e^{B_t^H} - 1\right) e^{rt+B_t^H}$$
$$= e^{rt} \left(1 - e^{2B_t^H} + 2e^{2B_t^H} - 2e^{B_t^H}\right) \qquad (4.6)$$
$$= e^{rt} \left(e^{B_t^H} - 1\right)^2.$$

Applying the chain rule to this, one gets

$$dX_t^\pi = re^{rt} \left(e^{B_t^H} - 1\right)^2 dt + 2e^{rt+B_t^H} \left(e^{B_t^H} - 1\right) \delta B_t^H. \qquad (4.7)$$

Looking at the second term of the sum, we get

$$2e^{rt+B_t^H} \left(e^{B_t^H} - 1\right) \delta B_t^H = \gamma_t S_t \, \delta B_t^H. \qquad (4.8)$$

The first term of this sum can be rewritten by

$$re^{rt} \left(e^{B_t^H} - 1\right)^2 dt = e^{2B_t^H} re^{rt} \, dt - 2re^{rt+B_t^H} \, dt + re^{rt} \, dt$$
$$= 2re^{rt+B_t^H} \, dt \left(e^{B_t^H} - 1\right) + re^{rt} \, dt \left(1 - e^{2B_t^H}\right) \quad (4.9)$$
$$= \gamma_t S_t r \, dt + \beta_t r B \, dt.$$

Combining (4.7)–(4.9), we get

$$dX_t^\pi = \gamma_t S_t \left(r \, dt + \delta B_t^H\right) + \beta_t r B \, dt$$
$$= \beta_t \, dA_t + \gamma_t \, dS_t.$$

Hence, the strategy $\pi_t = (\beta_t, \gamma_t)$ is self-financing. The initial capital needed is $X_0^\pi = 0$. Looking at (4.6), the resulting portfolio value is always nonnegative, more than this, it is almost surely positive. Consequently, the presented strategy is an arbitrage strategy.

Note that Shiryayev (1998) considers only the case of $H > 0.5$. By use of the generalized definition of pathwise integration by Bender (2003a), all results hold true also in the antipersistent case.

4.1.2 Arbitrage in the Continuous Time Setting Using Wick-Based Integration

As seen in the preceding subsection, the first results concerning financial market models based on fractional Brownian motion looked rather disillusioning. Nevertheless, it was still hoped to remedy the shortcomings of the suggested market setting. The research interest in this field was re-encouraged by the new insights in stochastic analysis mainly initiated by the work of Duncan et al. (2000). As introductorily mentioned in the preliminary chapter, they provided a stochastic integration calculus with respect to fractional Brownian motion that is based on the Wick product. This integration concept makes it possible to draw parallels to the well-known Itô calculus. Actually, by the work of Duncan et al. (2000) as well as the research outcome of the following years, many of the useful tools applied in the classical Markovian case could be translated to the fractional, Wick-calculus-based world. To name only the most important results, we mention a fractional Itô theorem, a fractional Girsanov theorem or a fractional Clark–Ocone formula (for a detailed survey see Bender (2003a)). As a consequence, efforts at deriving no-arbitrage-based valuation methods were reinforced. Of particular interest was one property of this integration calculus: The unconditional expectation of the stochastic integral $\int_t^T \sigma S(s) dB_s^H$ equals zero. This feature was at least more promising than the situation in the setup based on pathwise integration (see also the discussion in Chap. 2).

Despite the innovative stochastic calculus, things did not really change for the better. Already several years before, Delbaen and Schachermayer (1994) had proved a more general result holding for the continuous market models: Irrespective of the choice of integration theory, a weak form of arbitrage called free lunch with vanishing risk can only be excluded if and only if the underlying stock price process S is a semimartingale. It is easy though to verify that, due to their persistent character, processes driven by fractional Brownian motion are not semimartingales (for a motivating access to this topic, see also the discussion of the discrete framework in Chap. 3).

Perhaps, the existence of free lunch with vanishing risk might have been accepted from a puristic and formal point of view in favor of the applicability of a preference-free pricing approach using no arbitrage arguments. Yet, Cheridito (2003) succeeded to construct explicit arbitrage strategies both in the fractional Bachelier model and in the fractional Black–Scholes market no matter which integration theory—pathwise or Wick-based calculus—is used. The examples and results quoted by Cheridito (2003) must be regarded as conceptual and formal refinements of those by Rogers (1997), smoothing out the problems mentioned on the very spot.

Bender (2003a) gives a different proof which we will sketch in the following. It is similar to the one by Shiryayev (1998) which we presented for the pathwise integration case. Again, we have the familiar market setting with the bond following

$$dA_t = rA_t\, dt, \tag{4.10}$$

and the stock—in contrast to Shiryayev (1998) allowing for an arbitrary constant drift μ and volatility σ—following

$$dS_t = \mu S_t\, dt + \sigma S_t\, dB_t^H. \tag{4.11}$$

Now, the differential equation is interpreted in the Wick sense, yielding the explicit representations

$$A_t = e^{rt}$$

and

$$S_t = S_0 \exp\left(\mu t - \frac{1}{2}\sigma^2 t^{2H} + \sigma B_t^H\right).$$ (4.12)

The strategy $\pi_t = (\beta_t, \gamma_t)$ of dynamic bond and stock positions that we are interested in is defined by

$$\beta_t = 1 - \exp\left(-2rt + 2\mu t - \sigma^2 t^{2H} + 2\sigma B_t^H\right),$$ (4.13)

$$\gamma_t = 2S_0^{-1}\left(\exp\left(-rt + \mu t - \frac{1}{2}\sigma^2 t^{2H} + \sigma B_t^H\right) - 1\right).$$ (4.14)

Like in the proof for pathwise integration in Subsect. 4.1.1, one can now calculate the portfolio value $V_t^\pi = \beta_t A_t + \gamma_t S_t$ and obtains a quadratic and thereby strictly positive expression. Again, it is straightforward to prove that the portfolio is furthermore self-financing and hence admits a riskless gain. In other words, π is an arbitrage strategy. For the lengthy calculations, we refer to (Bender (2003a), p. 145–146).

4.2 Diverse Approaches to Exclude Arbitrage

Several modifications of the fractional market setting have been suggested to avoid the existence of arbitrage. In particular, the approaches by Hu and Øksendal (2003) and by Elliott and van der Hoek (2003) initiated an intense debate as to what extent the Wick product is suited for use in a financial context. We reconstruct this discussion. In the following subsections, we present further possibilities of how the absence of arbitrage can be ensured. By name, these will be the introduction of market imperfections as well as modifications of the underlying stochastic process.

4.2.1 Excluding Arbitrage by Extending the Wick Product on Financial Concepts

Actually, the above-mentioned statements Delbaen and Schachermayer (1994), Cheridito (2003) and Bender (2003a) hold true as long as the definitions of the fundamental concepts as arbitrage, self-financing properties and admissibility remain unchanged. Hence, concepts have been proposed to overcome the existing difficulties by modification of the underlying definitions, among them the approaches due to Hu and Øksendal (2003) and Elliott and van der Hoek (2003). They extend the idea of Wick calculus beyond integration theory and change the definitions of the portfolio value and/or the property of being self-financing, incorporating the Wick product. We examine these approaches as well as the related literature and critique in the following more closely. In order to be able to relate to the according discussion, we will briefly recall the approach by Hu and Øksendal (2003). For comparison, we will subsequently also sketch the main idea of the contemporaneous work of Elliott and van der Hoek (2003).

Both Hu and Øksendal (2003) p. 24 et seq., and Elliott and van der Hoek (2003) p. 320 et seq., deal with the fractional Brownian Black–Scholes market defined by (4.10) and (4.11). The differential equation of the stock is again interpreted in the Wick-based sense: that is, the stock price process has the explicit representation of (4.12).

Now, the crucial innovation comes into play: In the article by Hu and Øksendal (2003), the value process of the portfolio $\pi_t = (\beta_t, \gamma_t)$ is assumed to be given by the stochastic process

$$Z_t^\pi = \beta_t A_t + \gamma_t \diamond S_t. \tag{4.15}$$

Furthermore, the authors replace the definition of a portfolio to be self-financing by the following property: A portfolio is said to be Wick self-financing if its process satisfies

$$dZ_t^\pi = \beta_t \, dA_t + \gamma_t \diamond dS_t \tag{4.16}$$
$$:= \beta_t r A_t \, dt + \mu \gamma_t \diamond S_t \, dt + \sigma \gamma_t \diamond S_t \, dB_t^H, \tag{4.17}$$

where the stochastic differential is again interpreted in the Wick sense. Solving (4.15) for β_t, one obtains

$$\beta_t = \frac{Z_t^\pi - \gamma_t \diamond S_t}{A_t}. \tag{4.18}$$

If π is self-financing, (4.17) holds and one can substitute β_t by the preceding expression. Hence, the authors get

$$dZ_t^\pi = r Z_t^\pi \, dt + \sigma \gamma_t \diamond S_t \left(\frac{\mu - r}{\sigma} \, dt + dB_t^H \right). \tag{4.19}$$

Defining

$$\tilde{B}_t^H := \frac{\mu - r}{\sigma} t + B_t^H, \tag{4.20}$$

and applying the fractional Girsanov theorem (see Sect. 2.5), (4.19) can be rewritten once more by

$$dZ_t^\pi = rZ_t^\pi\, dt + \sigma\gamma_t \diamond S_t\, d\tilde{B}_t^H. \tag{4.21}$$

Here, $\tilde{B}_t^H$ is a fractional Brownian motion under the new probability measure which will be denoted by $\tilde{P}^H$ and which is determined by the Girsanov change of measure. Hu and Øksendal (2003) proceed by multiplying both sides of (4.21) by e^{-rt}. From the fractional version of Itô's Lemma we know that

$$d\left(e^{-rt}Z_t^\pi\right) = e^{-rt}\, dZ_t^\pi - re^{-rt}Z_t\, dt.$$

Hence, integrating both sides from 0 to T, the authors receive

$$e^{-rT}Z_T^\pi = Z_0^\pi + \int_0^T e^{-rt}\sigma\gamma_t \diamond S_t\, d\tilde{B}_t^H.$$

Taking expectations on both sides with respect to the new measure $\tilde{P}^H$, one ends up with the following equation:

$$E_{\tilde{P}^H}\left[e^{-rT}Z_T^\pi\right] = Z_0, \tag{4.22}$$

as the stochastic integral of Wick type has zero expectation. If the portfolio π was an arbitrage, the portfolio value in time T should be nonnegative for all states of nature. Moreover, it should be positive on a set of states with positive probability under the real measure P^H. As the measures are equivalent, this would immediately imply that the expectation on the left side of (4.22) would be positive. Then, Z_0 would also have to be positive which contradicts the condition of zero or negative initial investment. Therefore, π cannot be an arbitrage strategy. Consequently, there is no strategy making a riskless gain out of nothing: that is, the market setting excludes arbitrage.

Hu and Øksendal (2003) p. 26, furthermore succeed in showing that their kind of fractional Black–Scholes market is complete, implying that the martingale measure mentioned above is the only one. Pricing is then simply done by taking expectations with respect to this unique measure and discounting with the riskless interest rate. The formula the authors derive for a European call option at time 0 with strike K and maturity T is the following:

$$C_0^H = S_0 N\left(d_1^*\right) - Ke^{-rT}N\left(d_2^*\right),$$

where

$$d_1^* = \frac{\ln(\frac{S_0}{K}) + rT + \frac{1}{2}\sigma^2 T^{2H}}{\sigma T^H},$$

$$d_2^* = \frac{\ln(\frac{S_0}{K}) + rT - \frac{1}{2}\sigma^2 T^{2H}}{\sigma T^H}.$$

Obviously, for $H = \frac{1}{2}$, one gets the well-known Black–Scholes pricing formula.

The same result was subsequently derived by Elliott and van der Hoek (2003), however their approach is a little bit different from that by Hu and Øksendal (2003). While the definition of the portfolio value is still based on ordinary multiplication, that is

$$Z_t^\pi = \beta_t A_t + \gamma_t S_t, \tag{4.23}$$

the self-financing condition is said to be satisfied by portfolios π, for which the following property holds:

$$dZ_t^\pi = \beta_t \, dA_t + \gamma_t \diamond \left(\mu S_t \, dt + \sigma W_t^H \, dt \right). \tag{4.24}$$

Here, W_t^H is the fractional analogue of white noise, called fractional white noise. The corresponding fractional white noise calculus is also developed by the authors (see Elliott and van der Hoek (2003), p. 309 et seqq.). Again, the market becomes free of arbitrage and completeness can be shown. The option pricing formula is a slight generalization of the formula of Hu and Øksendal (2003) given above.

The conceptual innovation of extending the Wick product to the definitions of value process and/or self-financing condition initiated an intense debate. Before dwelling on it, we should mention a peculiarity inhering in the pricing formula itself. Although the formula provided by Hu and Øksendal (2003) looks quite promising at first glance, this is mainly due to the fact that it prices the call option at the special point in time $t = 0$. However, if the formula is generalized to an arbitrary current time t (as done in Necula (2002) and Elliott and van der Hoek (2003)), the terms T^{2H} and T^H are replaced by the expressions $\left(T^{2H} - t^{2H} \right)$ and $\left(T^H - t^H \right)$ respectively. This is somehow irritating, as in this case, the option value not only depends on time to maturity $(T - t)$ (which would be the case if we had $(T - t)^{2H}$), but is up to the current point in time t. This makes it necessary to determine the line of time absolutely and not only relatively and leaves the question unanswerable of how this should be done.

From a pure mathematical point of view, the approaches by Hu and Øksendal (2003) as well as by Elliott and van der Hoek (2003) are formally correct and accurate. Actually, the encouraging result of a fractional Black–Scholes market excluding arbitrage entailed further models based thereon (e.g. see Necula (2002) or Benth (2003)). However, severe critique arose concerning the economic meaning of Wick products that are used beyond pure integration theory. By name, one has to talk about the feasibility of Wick-based definitions of fundamental economic concepts like the portfolio value process and the property of being self-financing. The first ones to point at eventual problems of this kind were probably Sottinen and Valkeila (2003) who doubted the suitability of the Wick self-financing property when being concerned with economic questions. Bjork and Hult (2005) lately showed that extending the Wick product to the definitions of the portfolio value and the self-financing property indeed yields peculiar results. The most striking of them are highlighted in the following:

- The definition of the value process introduced by Hu and Øksendal (2003) contradicts economic intuition: Even if one knows the current realization of the stock process and the position held in stocks, this is not sufficient to calculate the realization of the portfolio value. This is due to the fact that the Wick product is a product of random variables that cannot be calculated pathwise. Consequently, in order to be able to calculate the current portfolio value at time t, it would be necessary to know the prospective holdings for all possible states of nature at this point in time: that is, also for those states of nature not being realized. This even leads to the somehow strange situation that the value of a portfolio holding nothing but a positive amount of a stock having a positive value, can nonetheless be negative (see Bjork and Hult (2005)).

- The self-financing condition stated by Elliott and van der Hoek (2003) also disagrees with the underlying economic meaning. It rules out strategies from being self-financing that obviously satisfy the basic property that no money is externally added or removed. Amongst others, simple buy-and-hold strategies are excluded. The exclusion of these strategies yet is shown to be necessary, as otherwise arbitrage possibilities would immediately occur. Bjork and Hult (2005) create such an arbitrage strategy explicitly (see Bjork and Hult (2005)).

It should be stressed that the critique by Bjork and Hult (2005) only addresses the extensions of the Wick product beyond the stochastic integral of the price process. The Wick-based definition of the latter however is not concerned and we will further use this integration concept for the reasons of unbiased mean and formal conformity with the Itô calculus mentioned above. Furthermore, it is worth mentioning that in another article, Øksendal suggested an economic interpretation of the Wick-based portfolio value introducing so-called market observers (see Øksendal (2006)). The dynamics of the process S_t are interpreted as the fundamental firm value. This firm value has to be distinguished carefully from the observable market price. The latter is assumed to be the outcome of a statistic test function applied to the distribution of the stochastic process. Likewise, the authors distinguish between trading strategies as a stochastic process and portfolio holdings which again are the result of the application of a statistic test function. The approach stems from quantum mechanics and tries to justify the Wick product to be the natural way of defining both stochastic integrals and the portfolio value process. Per se, the approach is coherent, yet the setting appears to be quite artificial (see e.g. Bender et al. (2006)).

Last but not least, we mention the recent work by Bender et al. (2006), who summarized the approaches presented above from a unifying angle. They show that all of the approaches can be viewed as restrictions of the class of trading strategies. The problem then turns into judging the feasibility of the respective restrictions (see Bender et al. (2006)).

4.2.2 Regularization of Fractional Brownian Motion

Other approaches redefine the setting by modifying the stochastic stock process. One of the suggested ways out is that of regularization, initiated by Rogers (1997): The basic idea is to replace the weighting kernel $\varphi(t) = (t)_{+}^{H-\frac{1}{2}}$ of the integral representation of fractional Brownian motion by a related one. The aim is, that the resulting stochastic process is close to fractional Brownian motion but behaves somehow better and becomes a semimartingale. In particular, it was aimed to preserve the main features like long-range dependence, approximately model the same moment properties and meantime smooth the behavior of the kernel for small arguments. The latter behavior could be identified to be a crucial reason why fractional Brownian motion cannot be a semimartingale. Rogers (1997) suggests the kernel

$$\varphi(t) = (a + t^2)^{\frac{2H-1}{4}},$$

with a small constant a. The resulting process is a semimartingale.

Cheridito (2001a) instead proposes a regularization replacing the weighting kernel in a small interval $[0, b]$ partially by a linear function (see Cheridito (2001a), p. 57 et seqq.). The author also broaches the issue of option pricing. Though the chosen regularization implies the existence of a unique martingale measure, the derived option prices heavily depend on the form of the linear function. Furthermore, one cannot state a clear relation between the chosen form of the kernel and the precision of the approximation of fractional Brownian motion. In other words, the same precision can be achieved by two different kernels yielding totally different option prices. So there seems to be no indication which kernel and at the same time which price should be ideally chosen (see Sottinen and Valkeila (2003), p. 14).

4.2.3 Mixed Fractional Brownian Motion

The approach of mixed fractional Brownian motion by Cheridito (2001b) stems from a similar motivation. The idea again is to modify the stochastic process of the stock price in order to get a semimartingale. The so-called mixed model combines fractional Brownian motion with a classical Brownian motion yielding the stock price dynamics

$$dS_t = \mu S_t \, dt + \epsilon S_t \, dB_t + \sigma S_t \, dB_t^H.$$

For Hurst parameters $H \in (\frac{3}{4}, 1)$, it is shown, that there is a unique martingale measure, as long as the processes B_t and B_t^H are independent. Clearly, by this model one can approximate the stock process of geometric fractional Brownian motion arbitrarily well choosing ϵ accordingly close to zero. Particularly, the combined process $\epsilon B_t + \sigma B_t^H$ is of course Gaussian and has zero mean, while the covariance satisfies

$$Cov\left(\epsilon B_t + \sigma B_t^H, \epsilon B_s + \sigma B_s^H\right) = \epsilon^2 \min(t, s) + Cov\left(B_t^H, B_s^H\right).$$

For a chosen value ϵ, one can price assets with respect to the unique martingale measure Q_ϵ and gets (at current time $t = 0$)

$$
\begin{aligned}
C_0(\epsilon) &= E_{Q_\epsilon}\left[\max\left(S_0 \exp\left(\mu T + \sigma(\epsilon B_T + B_T^H)\right) - e^{-rT}K, 0\right)\right] \\
&= BS\left(0, S_0, \sigma\epsilon\right),
\end{aligned}
$$

where $BS\left(0, S_0, \sigma\epsilon\right)$ denotes the Black–Scholes price of a call on a stock with initial price S_0 and volatility $\sigma\epsilon$ (see Cheridito (2001b)). We have a closer look at the consequences of this result in Sect. 4.3.

4.2.4 Market Imperfections

A proximate solution to avoid the existence of arbitrage is the introduction of market imperfections. Guasoni (2006) asserts proportional transaction costs. He succeeds at proving the absence of arbitrage within the fractional Brownian market (see Guasoni (2006), p. 577–580). However, the problem of option pricing remains unsolved and we further do not address ourselves to this approach.

An approach that we think more promising is the renouncement of continuous tradability. If we state that a single investor cannot proceed two consecutive transactions infinitesimally fast, the market becomes free of arbitrage (see Cheridito (2003)). However, the market then is dynamically incomplete. In the following sections we provide a detailed motivation of this restriction on tradability.

4.3 On the Non-compatibility of Fractional Brownian Motion and Continuous Tradability

The result of the preceding section concerning the mixed model is most interesting: As ϵ tends to zero, the mixed model approaches the fractional Brownian motion market, but at the same time, the option price tends to

$$C_0^* = \lim_{\epsilon \to 0} BS\left(0, S_0, \sigma \epsilon\right) = \max\left(S_0 - Ke^{-rT}, 0\right), \qquad (4.25)$$

that is, all randomness is eliminated. Cheridito (2001b) verbally explains this peculiarity by the possibility that traders can act infinitely fast and hence immediately exploit the predictability of the fractional Brownian motion mixed model. Thereby, they remove the random character by means of a suitable trading strategy (see Cheridito (2001b)).

We give a calculative example that supports this conjecture by putting the work of Sethi and Lehoczky (1981) into a fractional context. From this example we will draw important consequences concerning the renouncement of market completeness. Besides this, the respective notes present one more illustration of the special character of fractional Brownian motion as a process exhibiting predictability. We first recall the original findings of Sethi and Lehoczky (1981).

4.3.1 Itô and Stratonovich Formulations of the Classical Option Pricing Problem: The Work of Sethi and Lehoczky (1981)

The central statements of Sethi and Lehoczky (1981) are the following: The Black–Scholes option pricing formula can be derived both in a setting using Itô integrals and in a Stratonovich framework, if all differentials are carefully interpreted. The misinterpretation of the Stratonovich differential—or more precisely, the inaccurate mixture of formulation, on the one hand, and interpretation, on the other hand—however leads to a formula comparable to (4.25) (see Sethi and Lehoczky (1981), p. 352).

To be more precise: in the article by Sethi and Lehoczky (1981), the authors first start with the formulation of the problem in terms of Itô calculus. For sake of facility, we only denote the arguments of the processes, when introducing the latter or wherever needed to avoid misunderstandings. In all differential equations, we leave out the arguments in brackets. The risky process $S(t)$ of geometric Brownian motion is given by

$$(I) \quad dS = \mu S \, dt + \sigma S \, dB_t, \tag{4.26}$$

where the letter I in brackets anteceding the equation indicates that the equation is formulated in the Itô sense and B_t is classical Brownian motion. Let $C(t, S(t))$ denote the value of a European call option at time t. Using Itô's lemma, we get the dynamics of this derivative to satisfy

$$(I) \quad dC = \left[\frac{\partial C}{\partial t} + \mu \frac{\partial C}{\partial S} S + \frac{1}{2} \sigma^2 \frac{\partial^2 C}{\partial S^2} \right] dt + \sigma \frac{\partial C}{\partial S} S \, dB_t. \tag{4.27}$$

If one forms a dynamical portfolio $R(t, S_t)$ consisting of

- one unit of the option C,
- a short position on $\frac{\partial C}{\partial S}$ units of the stock S and
- a debt of $A(t, S)$ at the risk-free interest rate r,

this portfolio satisfies

$$(I) \quad dR = dC - \frac{\partial C}{\partial S} dS - Ar \, dt$$

$$= \left[\frac{\partial C}{\partial t} + \mu \frac{\partial C}{\partial S} S + \frac{1}{2} \sigma^2 \frac{\partial^2 C}{\partial S^2} S^2 \right] dt + \sigma \frac{\partial C}{\partial S} S \, dB_t$$

$$- \frac{\partial C}{\partial S} dS \left[\mu S \, dt + \sigma S \, dB_t \right] - Ar \, dt.$$

The stochastic part vanishes and if one adjusts the position A dynamically by

$$(I) \quad A(t, S) = \left(\frac{\partial C}{\partial t} + \frac{1}{2} \sigma^2 S^2 \frac{\partial^2 C}{\partial S^2} \right) / r,$$

one gets $dR = 0$. Obviously, the portfolio does not yield any return, hence its value itself must also be zero. Exploiting this, leads to the well-known differential equation

$$(I) \quad \frac{1}{2} \sigma^2 \frac{\partial^2 C}{\partial S^2} S^2 + r \frac{\partial C}{\partial S} S + \frac{\partial C}{\partial t} - rC = 0, \tag{4.28}$$

which has to be solved with respect to the boundary conditions

$$C(t, 0) = 0, \qquad\qquad \forall t \in [0, T] \tag{4.29}$$

$$\text{and} \quad C(T, S(T)) = \max[S(T) - K, 0]. \tag{4.30}$$

One obtains the classical Black–Scholes pricing formula.

We know from Sect. 2.5 that there is an easy link between the Itô integral and the Stratonovich integral. If one wants to describe the same market setting in terms of Stratonovich calculus, the equation of the basic risky asset has to be rewritten and (4.26) turns into the following (see Sethi and Lehoczky (1981), p. 351):

$$(S) \quad dS = \left(\mu - \frac{1}{2}\sigma^2\right) S\,dt + \sigma S\,dB_t, \tag{4.31}$$

where the letter S in brackets anteceding the equation indicates that the equation is formulated in the Stratonovich sense. Instead of Itô's lemma, we now apply the classical chain rule when deriving the dynamics of the derivative $C(t, S)$. We get

$$(S) \quad dC = \left[\frac{\partial C}{\partial t} + \left(\mu - \frac{1}{2}\sigma^2\right)\frac{\partial C}{\partial S}S\right] dt + \sigma\frac{\partial C}{\partial S}S\,dB_t. \tag{4.32}$$

Forming the same portfolio R as above with

$$(S) \quad dR = dC - \frac{\partial C}{\partial S}dS - Ar\,dt, \tag{4.33}$$

and inserting (4.31) and (4.32), yields

$$(S) \quad dR = \left[\frac{\partial C}{\partial t} + \left(\mu - \frac{1}{2}\sigma^2\right)\frac{\partial C}{\partial S}S\right] dt + \sigma\frac{\partial C}{\partial S}S\,dB_t$$
$$- \frac{\partial C}{\partial S}\left[\left(\mu - \frac{1}{2}\sigma^2\right)S\,dt + \sigma S\,dB_t\right] - Ar\,dt.$$

Again, the portfolio is locally riskless. In order to remove also the deterministic part, one has to set

$$(S) \quad A(t, S) = \frac{\partial C}{\partial t}/r.$$

With this, the value of the portfolio must also equal zero and Sethi and Lehoczky (1981) receive the following partial differential equation:

$$(S) \quad r\frac{\partial C}{\partial S}S + \frac{\partial C}{\partial t} - rC = 0.$$

Solving this with respect to the boundary conditions (4.29) and (4.30), the authors obtain

$$(S) \quad C(t, S(t)) = \max[S(t) - Ke^{-r(T-t)}, 0]. \tag{4.34}$$

This result looks surprising at first glance. The option value in time t is nothing but the difference between the current value of the stock and the discounted strike price or—if the latter difference is negative—at least zero. Apparently, in case of this interpretation of integration calculus, the randomness of the contract can be eliminated by a continuously-adapting hedging strategy. This hedging strategy exploits the predictability that may come into play when using Stratonovich integrals.

However, the problem can be solved by having a closer look at the derivation of the preceding formula. Sethi and Lehoczky (1981) show that if one has to decide whether to take Stratonovich or Itô integrals one has to take into account the respective implied economic meaning. In the upper example, it is crucial to give the correct interpretation concerning the dynamics of the process $\frac{\partial C}{\partial S} S$. In the case of classical Brownian motion, if formulated in terms of the Itô calculus, we call this product $Y(t, S(t))$ and it can be described by

$$(I) \quad dY = \frac{\partial C}{\partial S}(t, S(t)) \left(S(t+dt) - S(t) \right).$$

If the same process is denoted in Stratonovich formulation, we call it $Z(t, S(t))$ and it has to be interpreted as follows:

$$(S) \quad dZ = \frac{\partial C}{\partial S} \left(t, \frac{S(t) + S(t+dt)}{2} \right) \left(S(t+dt) - S(t) \right)$$

Note that these two equations describe different processes. The usage of the latter concept would assume that investors can adapt their portfolio holdings $\frac{\partial C}{\partial S}$ at the midpoint of an interval, but stand to benefit from the increment of the total interval. To rule out this sort of clairvoyance of investors, the Itô interpretation is the only feasible one. In this case, investors can only fix their portfolio holdings at the starting point of the incremental interval. However, if one wants to make use of a Stratonovich notation of $dY^{(I)}$, it is possible to translate the respective Itô term into this very notation and one obtains (see Sethi and Lehoczky (1981), p. 354):

$$(S) \quad dY = \frac{\partial C}{\partial S} S - \frac{1}{2}\sigma^2 \frac{\partial^2 C}{\partial S^2} dt$$
$$= \left[\frac{\partial C}{\partial S} \mu S - \frac{1}{2}\sigma^2 \frac{\partial^2 C}{\partial S^2} - \frac{1}{2}\sigma^2 \frac{\partial C}{\partial S} S \right] dt + \sigma \frac{\partial C}{\partial S} S \, dB_t.$$

Comparing this with dZ, we get aware of the additional term $-\frac{1}{2}\sigma^2 \frac{\partial^2 C}{\partial S^2} dt$. Inserting for the unspecified term $\frac{\partial C}{\partial S} S$ in (4.33) the correct interpretation $dY^{(S)}$, the portfolio value turns into

$$(S) \quad dR = \left[\frac{\partial C}{\partial t} + \left(\mu - \frac{1}{2}\sigma^2 \right) \frac{\partial C}{\partial S} S \right] dt + \sigma \frac{\partial C}{\partial S} S \, dB_t$$
$$- \frac{\partial C}{\partial S} \left[\left(\mu - \frac{1}{2}\sigma^2 \right) S \, dt + \sigma S \, dB_t \right] + \frac{1}{2}\sigma^2 \frac{\partial^2 C}{\partial S^2} dt - Ar \, dt.$$

Following the same steps as before, we now get

$$(S) \quad A(t, S) = \left[\frac{\partial C}{\partial t} + \frac{1}{2} \sigma^2 \frac{\partial^2 C}{\partial S^2} \right] / r,$$

and obtain the partial differential equation (4.28) which is the well-known result of Black, Scholes and Merton. Consequently, also for the option value we receive the familiar Black–Scholes pricing formula.

Let us summarize the main contribution of Sethi and Lehoczky (1981): For the interpretation of economic aspects like a trading strategy or a portfolio holding, a midpoint evaluation of Stratonovich type implies the ability to look into the future. Trading strategies could then be adjusted to the available information: To put it differently, at the time the trading strategy is fixed, the investor could anticipate parts of the evolution of the asset in question and thereby eliminate uncertainty. The resulting option pricing formula (4.34) accounts for this loss of randomness. Consequently, concerning the interpretation of the crucial terms, there is only one correct integration concept, the Itô calculus. However, it is always possible to translate an Itô type integral into Stratonovich notation. Hence, it is possible to formulate the option pricing problem both in terms of Itô and Stratonovich calculus. Once the very differentials are correctly interpreted, the chosen notation will not change the outcome of the calculations problem: Both notations yield the well-known option pricing formula.

4.3.2 Wick–Itô and Stratonovich Formulations of the Fractional Option Pricing Problem

We now reconsider the problem in a fractional context. In terms of Wick–Itô integration calculus, the stock price $S(t)$ exhibits the well-known dynamics of geometric fractional Brownian motion

$$(W) \quad dS = \mu S\, dt + \sigma S\, dB_t^H \tag{4.35}$$

The same trajectories are obtained by the following process written in Stratonovich notation:

$$(S) \quad dS = \left(\mu - H\sigma^2 t^{2H-1} \right) S\, dt + \sigma S\, dB_t^H. \tag{4.36}$$

The letters in brackets anteceding the equations stand for Wick–Itô or for fractional Stratonovich integration, respectively. Irrespective of the chosen integration concept, the time continuous riskless interest rate is r.

We first look at the Stratonovich case. When calculating the dynamics of a call option $C(t, S_t)$, we hence have to apply the chain rule and get

$$(S) \quad dC = \frac{\partial C}{\partial t} dt + \frac{\partial C}{\partial S} dS$$

$$= \left(\frac{\partial C}{\partial t} + \frac{\partial C}{\partial S} (\mu - H\sigma^2 t^{2H-1})S \right) dt + \frac{\partial C}{\partial S} \sigma S \, dB_t^H. \quad (4.37)$$

Like Sethi and Lehoczky (1981) we consider a portfolio $R(t, S_t)$ consisting of

- one unit of the option C,
- a short position on $\frac{\partial C}{\partial S}$ units of the stock S and
- a debt of $A(t, S)$ at the risk-free interest rate r.

With $Y(t, S) = \frac{\partial C}{\partial S} S$, the corresponding dynamics of the portfolio R are

$$(S) \quad dR = dC - dY - Ar \, dt. \quad (4.38)$$

In the classical Brownian setting, it was crucial to ascribe the correct meaning to dY. In particular, it was important to avoid a midpoint evaluation that contradicts economic intuition (see Sethi and Lehoczky (1981)). Here, the specific character of the fractional setting comes into play: For $H \neq \frac{1}{2}$, the fractional Stratonovich integral is independent of the point the evaluation of the integrand takes place (see Duncan et al. (2000)). Hence, the desired interpretation

$$dY = \frac{\partial C}{\partial S}(t, S(t))[S(t + dt) - S(t)]$$

is already of Stratonovich type and necessitates no further translation. Consequently, no additional term occurs and one can simply write

$$(S) \quad dY = \frac{\partial C}{\partial S} dS. \quad (4.39)$$

Inserting (4.37) and (4.39) into (4.38), one gets

$$(S) \quad dR = \left(\frac{\partial C}{\partial t} + \frac{\partial C}{\partial S} (\mu - H\sigma^2 t^{2H-1})S \right) dt + \frac{\partial C}{\partial S} \sigma S \, dB_t^H$$

$$- \frac{\partial C}{\partial S} \left((\mu - H\sigma^2 t^{2H-1})S \, dt + \sigma S \, dB_t^H \right) - Ar \, dt$$

$$= \left(\frac{\partial C}{\partial t} - Ar \right) dt. \quad (4.40)$$

The portfolio therefore is locally riskless. With

$$A(t, S) = \frac{\partial C}{\partial t} / r,$$

we get a local return of zero. One can conclude that in this case the value of the portfolio also has to be zero yielding

$$C - \frac{\partial C}{\partial S}S - A = 0,$$

or, respectively

$$\frac{\partial C}{\partial t} - rC - rS\frac{\partial C}{\partial S} = 0.$$

In combination with the boundary conditions for the European call option, one gets the solution

$$C(t, S(t)) = \max\left(S(t) - Ke^{-r(T-t)}, 0\right),$$

which is the deterministic solution of (4.34).

We now address ourselves to the case of Wick-based integration characterized by (4.35). With regard to the dynamics of the call option $C(t, S_t)$, we apply the fractional Itô formula of Duncan et al. (2000) and obtain

$$(W) \quad dC = \frac{\partial C}{\partial t}\,dt + \frac{\partial C}{\partial S}\mu S\,dt + \frac{\partial C}{\partial S}\sigma S\,dB_t^H + \frac{\partial^2 C}{\partial S^2}\sigma^2 S^2(t^{2H-1})\,dt. \quad (4.41)$$

To make the notation clearer, we rewrite the third term of the right hand side by

$$(W) \quad \frac{\partial C}{\partial S}\sigma S\,dB_t^H = \left(\frac{\partial C}{\partial S}\sigma S\right) \diamond dB_t^H. \quad (4.42)$$

We consider the same portfolio R as before. With $Y(t, S) = \frac{\partial C}{\partial S}S$, the corresponding dynamics of the portfolio R are

$$(W) \quad dR = dC - dY - Br\,dt. \quad (4.43)$$

Again, we have to take a closer look at the process Y and its dynamics. We have

$$(W) \quad dY = \frac{\partial C}{\partial S}\,dS$$
$$= \frac{\partial C}{\partial S}\mu S\,dt + \frac{\partial C}{\partial S}\sigma S\,dB_t^H. \quad (4.44)$$

The second part of the term has to be interpreted with care. Necula (2002) proposes to take

$$\frac{\partial C}{\partial S}\sigma S\,dB_t^H = \left(\frac{\partial C}{\partial S}S\sigma\right) \diamond dB_t^H,$$

which leads to a very promising differential equation (see Necula (2002), p. 15 et seq.). Using the according boundary conditions of the European call, the resulting formula equals the results of Elliott and van der Hoek (2003) as well as Hu and Øksendal (2003) mentioned above. Actually, the way of interpreting dY used by Necula (2002) corresponds to the modified definition of the portfolio using Wick products chosen by Hu and Øksendal (2003) (see Sect. 4.2). However, a correct economic interpretation of dY should use ordinary multiplication between the portfolio holding on the one hand and the asset value on the other hand, that is

$$(W) \quad \frac{\partial C}{\partial S} \sigma S \, dB_t^H = \frac{\partial C}{\partial S} \left(\sigma S \diamond dB_t^H \right). \tag{4.45}$$

Taking into account the relations between Wick products and ordinary products, we have

$$(W) \quad \frac{\partial C}{\partial S} \left(\sigma S \diamond dB_t^H \right) = \left(\frac{\partial C}{\partial S} \sigma S \right) \diamond dB_t^H + \frac{\partial^2 C}{\partial S^2} \sigma^2 S^2 (t^{2H-1}) \, dt. \tag{4.46}$$

If we combine (4.42)–(4.46), we receive the portfolio dynamics

$$
\begin{aligned}
(W) \quad dR ={}& \frac{\partial C}{\partial t} \, dt + \frac{\partial C}{\partial S} \mu S \, dt + \left(\frac{\partial C}{\partial S} \sigma S \right) \diamond dB_t^H + \frac{\partial^2 C}{\partial S^2} \sigma^2 S^2 (t^{2H-1}) \, dt \\
&- \left(\frac{\partial C}{\partial S} \mu S \, dt + \left(\frac{\partial C}{\partial S} \sigma S \right) \diamond dB_t^H + \frac{\partial^2 C}{\partial S^2} \sigma^2 S^2 (t^{2H-1}) \, dt \right) \\
&- Ar \, dt \\
={}& \left(\frac{\partial C}{\partial t} - Ar \right) dt.
\end{aligned}
$$

Again, the portfolio return is deterministic. Choosing as above

$$A(t, S) = \frac{\partial C}{\partial t} / r,$$

its return vanishes. Hence, the portfolio value should also equal zero, implying

$$(W) \quad \frac{\partial C}{\partial t} - rC - rS \frac{\partial C}{\partial S} = 0.$$

Consequently, the Wick–Itô notation of the problem yields the same solution as in the Stratonovich setting before, where we got (4.34). Recall, that this solution was also obtained in the previous section when discussing the limit case of the mixed fractional model by Cheridito (2001b).

Obviously, the example of Sethi and Lehoczky (1981) has been inverted in the fractional framework. The proper treatment of the differentials leads for both integration concepts to the deterministic solution, the non-deterministic solution of Necula (2002) is only due to a misleading interpretation of the differential dY. To put it differently: In the classical Brownian model, there is no predictability no matter which integration calculus is used. On the other hand, in the fractional model, predictions are always possible. Again the choice of integration calculus does not affect the result. By allowing for dynamic and infinitely-fast adapting strategies (in other words, assuming continuous tradability), the predictive nature of fractional Brownian motion can be exploited and uncertainty can be eliminated. Most plausibly, in a world like this, option values are equal to their discounted inner value.

4.4 Renouncement of Continuous Tradability, Exclusion of Arbitrage and Transition to Preference-Based Pricing

The given example is one more indication that fractional Brownian markets exhibit strange features if dynamical completeness of the market is assumed. The proximate consequence of giving up this market completeness will now be further pursued.

Cheridito (2003) takes into account the problems arising in fractional Brownian markets if one allows for continuous tradability and hence proposes a modification of the framework: He shows, that—when postulating the existence of an arbitrarily small minimal amount of time that must lie between two consecutive transactions—all kinds of arbitrage opportunities can be excluded (see Cheridito (2003), p. 549 et seq.). This is exactly the same argument we applied in the discrete time setting of binomial trees. Again, the central assertion is that traders can be arbitrarily fast, yet, not as fast as the market. Giving up continuous tradability, we thus obtain a reasonable financial model where no arbitrage occurs.

While this assumption of non-continuous trading strategies does not seem to be too restrictive when thinking of real financial markets, it entails one problem: Though having excluded arbitrage, the traditional no arbitrage option pricing approaches continue to fail, as now the possibility of a continuous adjustment of the replicating portfolio is no longer given. Dynamical hedging and replication methods are no longer available to derive prices of derivative assets. The consequence to which one may come, is to abstain from this kind of setting (see Bender et al. (2006)).

However, there are still other ways to derive prices in a financial market that is free of arbitrage. From the mathematical point of view, no arbitrage ensures the existence of an equivalent pricing measure. Yet, in lack of completeness of the market, this measure is not unique. The valuation problem therefore reduces to finding the most plausible one. On the other hand, economic theory provides us with equilibrium theory. From this point of view, the task is to find a suitable equilibrium condition. However, in both cases we will have to leave the world of preference-free pricing and introduce risk preferences instead. In the following chapter, we will seize the second of these two possibilities and focus on an equilibrium pricing approach.

Chapter 5
Risk Preference-Based Option Pricing in a Continuous Time Fractional Brownian Market

5.1 Motivation and Setup of the Model

After the success of risk-neutral valuation in the Markovian models by Black, Scholes and Merton (see Black and Scholes (1973), Merton (1973)), it was aspired to extend the famous option pricing formula for use in a fractional context. In the course of the last few years however, it turned out that no arbitrage pricing methods could not be sensibly applied within the fractional market model (see e.g. Rogers (1997) and Bjork and Hult (2005)).

In this chapter which further develops a preceding joint article of Rostek and Schöbel (2006), we look at a market where randomness is driven by fractional Brownian motion. While the price process evolves continuously, we will assume that a single investor faces certain restrictions concerning the speed of his transactions. In the sense of Cheridito (2003), we introduce a minimal amount of time—which can be arbitrarily small—between two consecutive transactions by one and the same investor. The idea behind this is that the great multitude of investors in the market ensures that there are other transactions in-between. In other words, this assumption restricts a single investor from being as fast as the market as a whole which evolves continuously. This restriction is sufficient to exclude arbitrage in the fractional Brownian market (see Cheridito (2003)).

Note that, as we give up on continuous tradability, the well-known no arbitrage pricing approach based on dynamical hedging arguments is unsuitable within this modified framework. This problem is solved the most naturally by introducing risk preferences (see Brennan (1979)). Concerning these risk preferences, the market has to satisfy a basic equilibrium condition which we will investigate. In the special case of risk-neutral investors, the option pricing problem will prove to be the calculation of the discounted conditional mean of the relevant payoff.

S. Rostek, *Option Pricing in Fractional Brownian Markets.*
Lecture Notes in Economics and Mathematical Systems.
© Springer-Verlag Berlin Heidelberg 2009

Further advantages of a transition to a preference based pricing approach will turn out to be the following: The use of conditional expectation in its traditional sense will make it possible to demonstrate the problems arising in valuation models when dealing with path-dependent processes. Moreover, advances in stochastic analysis will be used to plausibly illustrate the features of fractional Brownian motion and to make fractional option pricing comparable to the classical Brownian model. In particular, the consequences of the existence of long-range dependence on option pricing should be clarified.

Throughout this chapter, we assume a fractional Brownian market where we have two basic assets, a bond without any risk

$$A_t = A_0 \exp(rt),$$

as well as a risky stock

$$S_t = S_0 \exp\left(\mu t - \sigma^2 t^{2H} + \sigma B_t^H\right). \tag{5.1}$$

The coefficients r, μ, σ are assumed to be constants symbolizing the riskless interest rate, the drift of the stock and its volatility, respectively.

Concerning the stochastic calculus of fractional Brownian motion, Duncan et al. (2000) introduced a fractional Itô theorem. Due to its importance, we briefly recall this theorem here in a generalized version. If S_t is a geometric fractional Brownian motion as above and if $F(t, S_t)$ is once continuously differentiable with respect to t and twice with respect to S_t, then under additional regularity conditions one gets (see Bender (2003a)):

$$F(T, S_T) = F(t, S_t) + \int_t^T \frac{\partial}{\partial s} F(s, S_s)\, ds + \int_t^T \frac{\partial}{\partial x} F(s, S_s) \mu S_s\, ds$$
$$+ \sigma \int_t^T \frac{\partial}{\partial x} F(s, S_s) S_s\, dB_s^H + H\sigma^2 \int_t^T s^{2H-1} \frac{\partial^2}{\partial x^2} F(s, S_s) S_s^2\, ds. \tag{5.2}$$

Note that for the limit $H \to \frac{1}{2}$ the formula looks quite familiar to us, as we obtain

$$F(T, S_T) = F(t, S_t) + \int_t^T \frac{\partial}{\partial s} F(s, S_s)\, ds + \int_t^T \frac{\partial}{\partial x} F(s, S_s) \mu S_s\, ds$$
$$+ \sigma \int_t^T \frac{\partial}{\partial x} F(s, S_s) S_s\, dB_s + \frac{1}{2}\sigma^2 \int_t^T \frac{\partial^2}{\partial x^2} F(s, S_s) S_s^2\, ds, \tag{5.3}$$

which is the classical Itô theorem. From (5.2), it is easy to verify a fractional version of the so-called Doléans–Dade identity: Take $F(t, S_t) = S_t$ and apply formula (5.2). This yields

$$S_T = S_t + \int_t^T \mu S_s \, ds + \int_t^T \sigma S_s \, dB_s^H.$$

Consequently, we can also describe the two basic assets by differential equations, the riskless asset satisfying

$$dA_t = rA_t \, dt,$$

and the risky asset following

$$dS_t = \mu S_t \, dt + \sigma S_t \, dB_t^H.$$

Obviously, the model shown is a generalization of the seminal model by Black, Scholes and Merton (see Black and Scholes (1973), Merton (1973)). The only difference is that the diffusion term of classical Brownian motion is substituted by fractional Brownian motion including the classical case for $H = \frac{1}{2}$.

The rest of the chapter is organized as follows: In Sect. 5.2 we describe the conditionality of distributional forecasts, in particular, we will recall and interpret the results by Gripenberg and Norros (1996) and Nuzman and Poor (2000). Subsequently, as one of our central mathematical tools, a conditional fractional Itô theorem is derived. Section 5.3 will be devoted to this. Furthermore, we will focus in Sect. 5.4 on the risk preference based option pricing approach exemplified by the assumption of risk-neutral market participants. Particular focus on a two-time model will be made, whereby the equilibrium condition we introduce only takes into account present time t and maturity T. We will derive pricing formulae as well as further results and interpret them. Based on this, we stress the parallels between option pricing with geometric fractional Brownian motion and the classical Black–Scholes diffusion as underlying process, respectively.

5.2 The Conditional Distribution of Fractional Brownian Motion

5.2.1 Prediction Based on an Infinite Knowledge About the Past

In the first part of this section we focus on the distribution of fractional Brownian motion given all information concerning the history of the path. In particular, we look at $E[B_T^H | \mathfrak{F}_t^H], T > t$, where $\mathfrak{F}_t^H = \sigma(B_s^H, s \leq t)$ is the σ-field generated by all $B_s^H, s \leq t$. In the first instance, $E[B_T^H | \mathfrak{F}_t^H], T > t$ is a random variable, a coarsening of B_T^H, yielding in each case the expected value over all $\omega \in \Omega$ having the same path on $(-\infty, t]$. Knowing this kind of equivalence class $[\omega_1]_t = \{\omega \in \Omega | B_s^H(\omega) = B_s^H(\omega_1), \forall s \in (-\infty, t]\}$ from past observation, we will be able to conclude that the distribution of future realizations is again normal. Furthermore, we will be able to specify the distribution by use of the available information. We will observe an adjustment of the expected value as well as a variance reduction.

In the first step, we provide a representation formula for conditional expectation. Let $B_s^H, s \in \mathbb{R}$ be a fractional Brownian motion with $0 < H < 1$. Then, for each $T > t > 0$, the conditional expectation of B_T^H based on $\mathfrak{F}_t^H$ can be represented by:

$$\hat{B}_{T,t}^H = E\left[B_T^H | \mathfrak{F}_t^H\right] = B_t^H + (T-t)^{H+\frac{1}{2}} \int_{-\infty}^t g(T, t, s)\, ds, \qquad (5.4)$$

where

$$g(T, t, s) = \frac{\sin(\pi(H - \frac{1}{2}))(B_s^H - B_t^H)}{\pi(t-s)^{H+\frac{1}{2}}(T-s)}. \qquad (5.5)$$

The result is due to Nuzman and Poor (2000) and is an extension of the result by Gripenberg and Norros (1996) who proved the theorem for the case $t = 0$. Note that for technical reasons we translated the formula of Nuzman and Poor (2000) into the original notation of Gripenberg and Norros (1996). The proof uses both the self-similarity and the Gaussian character of fractional Brownianmotion. For $H > \frac{1}{2}$, an alternative representation can be given:

$$\hat{B}_{T,t}^H = E\left[B_T^H | \mathfrak{F}_t^H\right] = B_t^H + \int_{-\infty}^t g(T-t, s-t) dB_s^H, \qquad (5.6)$$

where

$$g(v, w) = \frac{\sin(\pi(H - \frac{1}{2}))}{\pi}(-w)^{-H+\frac{1}{2}} \int_0^v \frac{x^{H-\frac{1}{2}}}{x - w} dx \qquad (5.7)$$

$$= \frac{\sin(\pi(H - \frac{1}{2}))}{\pi}\left(\frac{1}{H - \frac{1}{2}}\left(\frac{-w}{v}\right)^{-H+\frac{1}{2}} - \beta_{v/(v-w)}\left(H - \frac{1}{2}, \frac{3}{2} - H\right)\right),$$

and $\beta.(\cdot, \cdot)$ is the incomplete Beta function. In both representations the conditional expectation of fractional Brownian motion consists of two parts. The first part is the current value of the process. In the case of a martingale this would already be the best prediction of the future, and indeed one can easily verify that for $H = \frac{1}{2}$ the second part vanishes. For all other Hurst parameters however, one gets an additional random term accounting for the historical evolution of the process. In the first representation (5.6) the randomness is hidden in the integrand $g(T, t, s)$. Although, at first glance, the second representation (5.7) looks more complicated, it is better suited to feed intuition. Here, the randomness stems from the integrator. To be more precise, the integration is done with respect to all historical fractional increments dB_s^H, from minus infinity to the current point in time. Hence, the additional term is calculated along the observed part of the path.

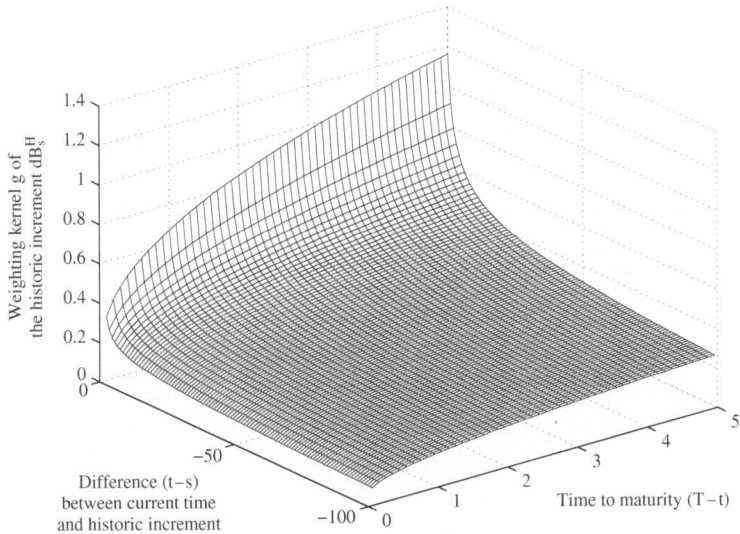

Fig. 5.1 Shape of the weighting kernel g of the conditional mean for $H = 0.8$

We take a closer look at the function $g(T, t, s)$. From the representation (5.7), one can see that the relation between the weighting kernel g and time to maturity $\tau = T - t$, is of order $H - \frac{1}{2}$. This implies a concave relation between τ and the conditional mean for the persistent case (see Fig. 5.1). Accordingly, one obtains a convex curve in the case of antipersistence. Concerning the distance of historic increments to current time, Fig. 5.1 illustrates that the most recent realizations have the greatest influence on the conditional mean. This is also true for the antipersistent case.

Let us for an instance fix current time t. We define equivalence classes by grouping paths $\omega_1, \omega_2, \ldots$ having identical trajectories up to time t into one class $[\omega_1]_t$. This means, although we cannot totally identify a particular path, we can at least determine its equivalence class by observing the historical part of the path.

For prediction purposes we are interested in the conditional distribution of B_T^H within its observed equivalence class. Let ω_1 be a representative of this equivalence class. The conditional distribution of B_T^H based on the observation $[\omega_1]_t$ is normal with the following moments:

$$E[B_T^H|\mathfrak{F}_t^H](\omega_1) = B_t^H(\omega_1) + \int_{-\infty}^t g(T, t, s)(\omega_1)\, ds := B_t^H(\omega_1) + \hat{\mu}_{T,t},$$

(5.8)

$$Var\left[B_T^H|\mathfrak{F}_t^H\right](\omega_1) = E\left[\left(B_T^H - \hat{B}_{T,t}^H\right)^2|\mathfrak{F}_t^H\right](\omega_1) = \rho_H(T-t)^{2H} := \hat{\sigma}_{T,t}^2,$$

(5.9)

with

$$\rho_H = \frac{\sin(\pi(H - \frac{1}{2}))}{\pi(H - \frac{1}{2})} \frac{\Gamma(\frac{3}{2} - H)^2}{\Gamma(2 - 2H)}.$$

We give a proof of this result. The normality of the conditional distribution is an immediate consequence of the Gaussian character of the process B_t^H. It is well known that Gaussian processes like multivariate normal distributions assure the normality of all kinds of conditional densities. Intuitively, the mean of the conditional distribution should be defined by $\int_{\omega \in [\omega_1]_t} B_T^H(\omega) d\hat{P}(\omega)$, where $\hat{P}(\omega) = \frac{P(\omega)}{P([\omega_1]_t)}$ is the conditional probability of ω. The characterization of the conditional mean given above then easily follows from (5.4)–(5.5) and the fact that the conditional expectation by definition satisfies:

$$\int_{\omega \in [\omega_1]_t} B_T^H(\omega) dP(\omega) = \int_{\omega \in [\omega_1]_t} \hat{B}_T^H(\omega) dP(\omega),$$

as $[\omega_1]_t \in \mathfrak{F}_t^H$. $\hat{B}_T^H$ being constant on $[\omega_1]_t$ we can rewrite this by

$$\int_{\omega \in [\omega_1]_t} B_T^H(\omega) dP(\omega) = \hat{B}_T^H(\omega_1) P([\omega_1]_t),$$

or, solving for $\hat{B}_T^H$,

$$\hat{B}_T^H(\omega_1) = \int_{\omega \in [\omega_1]_t} B_T^H(\omega) d\left(\frac{P(\omega)}{P([\omega_1]_t)}\right) = \int_{\omega \in [\omega_1]_t} B_T^H(\omega) d\hat{P}(\omega).$$

Respectively, the conditional variance should be defined by

$$\hat{\sigma}_{T,t}^2 = \int_{\omega \in [\omega_1]_t} \left[B_T^H(\omega) - \hat{B}_T^H(\omega)\right]^2 d\hat{P}(\omega),$$

which can be rewritten—applying the same argument as above—by

$$\hat{\sigma}_{T,t}^2 = E\left[(B_T^H - \hat{B}_T^H(\omega_1))^2|\mathfrak{F}_t^H\right](\omega_1).$$

But $\hat{B}_T^H$ is the orthogonal projection of B_T^H on the span of $\{B_s^H, s \leq t\}$. So, the co-projection $(B_T^H - \hat{B}_T^H)$ or $((B_T^H - B_t^H) - \hat{\mu}_{T,t})$, respectively, as well as the squared terms are orthogonal to and therefore independent of $\{B_s^H, s \leq t\}$. Hence, the conditional expectation $E\left[(B_T^H - \hat{B}_T^H(\omega_1))^2|\mathfrak{F}_t^H\right]$ is non-random. Consequently, we can omit the argument ω_1 in the following, add expectation operators and write:

$$\hat{\sigma}_{T,t}^2 = E\left[(B_T^H - \hat{B}_T^H)^2|\mathfrak{F}_t^H\right] = E\left(E\left[((B_T^H - B_t^H) - \hat{\mu}_{T,t})^2|\mathfrak{F}_t^H\right]\right)$$
$$= E\left(E\left[(B_T^H - B_t^H)^2|\mathfrak{F}_t^H\right] - 2E\left[(B_T^H - B_t^H)\hat{\mu}_{T,t}|\mathfrak{F}_t^H\right] + E\left[\hat{\mu}_{T,t}^2|\mathfrak{F}_t^H\right]\right)$$
$$= E(B_T^H - B_t^H)^2 - 2E(\hat{\mu}_{T,t})^2 + E(\hat{\mu}_{T,t})^2 = E(B_T^H - B_t^H)^2 - E(\hat{\mu}_{T,t})^2.$$

We now look at

$$E(\hat{\mu}_{T,t})^2 = E\left(\int_{-\infty}^t g\left((T-t),(s-t)\right) dB_s^H\right)^2$$
$$= \int_{-\infty}^t \int_{-\infty}^t g\left((T-t),(v-t)\right) g\left((T-t),(w-t)\right) \phi_H(v,w)dvdw$$
$$= \int_0^\infty \int_0^\infty g\left((T-t),(-x)\right) g\left((T-t),(-y)\right) \phi_H(x,y)dxdy$$
$$= (T-t)^{2H}(1-\rho_H),$$

where $\phi_H(a,b) = H(2H-1)|a-b|^{2H-2}$ and where we used Proposition 2.2 of Gripenberg and Norros (1996) and then substituted $x = t_0 - v$ and $y = t_0 - w$. The correctness of the last equality is carried out in the proof of Corollary 3.2 of Gripenberg and Norros (1996) where we refer to for more details. With that and

$$E((B_T^H - B_t^H)^2) = E(B_T^H)^2 - 2E(B_T^H B_t^H) + E(B_t^H)^2$$
$$= T^{2H} - (T^{2H} + t^{2H} - (T-t)^{2H}) + t^{2H} = (T-t)^{2H},$$

we get

$$\hat{\sigma}_{T,t}^2 = (T-t)^{2H} - (T-t)^{2H}(1-\rho_H) = \rho_H(T-t)^{2H},$$

which completes the proof.

Obviously, the conditional second moment only depends on the forecasting horizon $\tau = T - t$ and the Hurst parameter H, but not on the realized path. The representation of the conditional variance consists of two factors, where ρ_H only depends on the Hurst parameter. Figure 5.2 depicts its shape. The curve is always between zero and one, that is, the unconditional variance $(T-t)^{2H}$ is reduced by ρ_H. For $H = \frac{1}{2}$, the factor is maximally large, namely it equals one. In this case, there is no difference between unconditional and conditional variance, as history plays no role in the classical Brownian motion based setting. Another interesting issue can be observed if H tends to one. In the case of total persistence the future can be extrapolated exactly by knowing the historical data. Hence uncertainty vanishes and the factor ρ_H and consequently also the conditional variance tend towards zero. On the other hand, the maximal level of antipersistence (as $H \to 0$) does not eliminate all randomness. Historic information only halves the variance.

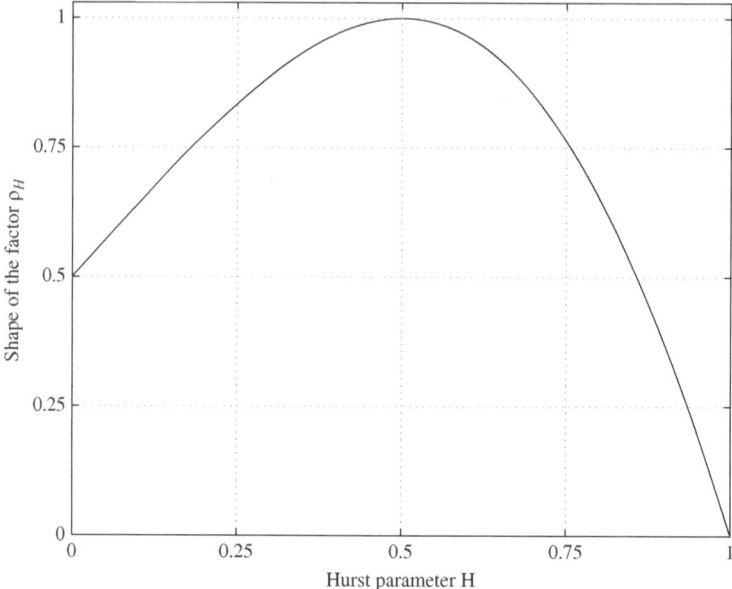

Fig. 5.2 Shape of the factor ρ_H

The second factor of $\hat{\sigma}_{T,t}^2$ is the unconditional variance $(T-t)^{2H}$. The relation between time to maturity and the variance hence can either exhibit a convex or a concave shape, depending on the Hurst parameter H (see Fig. 5.3). For $H = \frac{1}{2}$, the variance increases linearly with time to maturity.

Recall that while the conditional variance only depends on H and time to maturity, the conditional mean is really path-dependent and has to be calculated by means of (5.4) and (5.5) which entails evaluating the past. However, it seems to be quite difficult to make observations into an infinite past. In the rest of this chapter we focus on a finite observation interval.

5.2.2 Prediction Based on a Partial Knowledge About the Past

For practical purposes it is desirable to make predictions that are based on only a part of the past and to go back only to a finite point in time $t - a$. That is, we restrict ourselves to a finite observation interval of length a and

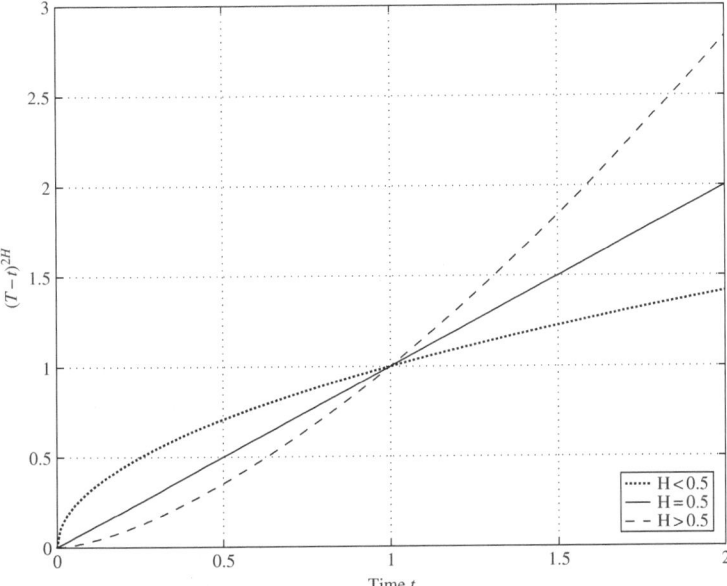

Fig. 5.3 Shape of the factor $(T-t)^{2H}$

focus on the distribution of B_T^H conditional on $\mathfrak{F}_{t,a}^H = \sigma(B_s^H, t-a \leq s \leq t)$, which is the σ-field generated by all $B_s^H, t-a \leq s \leq t$.

We state the according result concerning this kind of conditional expectation which we denote by $\hat{B}_{T,t,a}^H$. For all $0 < H < 1$ and all $T, t, a > 0$, the conditional expectation of B_T^H based on $\mathfrak{F}_{t,a}^H$ can be represented as follows (Nuzman and Poor (2000)):

$$\hat{B}_{T,t,a}^H = E[B_T^H | \mathfrak{F}_{t,a}^H] = \int_{t-a}^{t} g_a(T-t, t-s) B_s^H \, ds, \qquad (5.10)$$

where

$$g_a(x,y) = a^{-1} \left(\frac{y}{a}\right)^{-H-\frac{1}{2}} \left(1 - \frac{y}{a}\right)^{-H-\frac{1}{2}} \qquad (5.11)$$

$$\times \left[\left(\frac{1}{2} - H\right) \beta_{x/(x+1)} \left(H + \frac{1}{2}, 1-2H\right) + \frac{x^{H+\frac{1}{2}}(1+x)^{H-\frac{1}{2}}(1-y/a)}{x+y/a} \right].$$

Again, we can derive statements concerning the conditional distribution of fractional Brownian motion, in this instance based on limited knowledge about the past. The latter is expressed by the restriction to the equivalence class $[\omega_1]_t^a = \{\omega \in \Omega | B_s^H(\omega) = B_s^H(\omega_1), \forall t-a \leq s \leq t\}$: The conditional distribution of B_T^H based on the observation $[\omega_1]_t^a$ is normal with the following moments:

$$\hat{\mu}_{T,t,a} = E[B_T^H | \mathfrak{F}_{t,a}^H](\omega_1) = \int_{t-a}^{t} g_a(T-t, t-s) B_s^H(\omega_1) \, ds, \qquad (5.12)$$

$$\hat{\sigma}_{T,t,a}^2 = Var\left[B_T^H | \mathfrak{F}_{t,a}^H \right](\omega_1) := E\left[(B_T^H - \hat{B}_{T,t,a}^H)^2 | \mathfrak{F}_{t,a}^H \right](\omega_1)$$

$$= (T-t)^{2H}(1 - \rho_{H,a}), \qquad (5.13)$$

with

$$\rho_{H,a} := 1 - H \int_0^{\frac{a}{T-t}} g_{\frac{a}{T-t}}(1, -s) \left((1+s)^{2H-1} - s^{2H-1} \right) ds.$$

For $H > \frac{1}{2}$, there is again a more accessible representation of the conditional mean (see Gripenberg and Norros (1996)):

$$\hat{B}_{T,t,a}^H = B_t^H + \int_{t-a}^{t} g_a^*(T-t, t-s) dB_s^H, \qquad (5.14)$$

where

$$g_a^*(u, v) = \frac{\sin(\pi(H - \frac{1}{2}))}{\pi} (-v)^{-H+\frac{1}{2}}(a+v)^{-H+\frac{1}{2}} \int_0^u \frac{x^{H-\frac{1}{2}}(x+a)^{H-\frac{1}{2}}}{x - v} dx.$$

Formula (5.14) shows that the prediction formula—like in the case of infinite information—consists of the current value plus a correction term that is calculated by an integral over the observation interval. The proofs of the formulae (5.10)–(5.14) can be seen in Nuzman and Poor (2000) for the representation of the conditional mean and in Gripenberg and Norros (1996) for the result concerning the conditional variance. In order to avoid irritations, we adapted the results to our notation of the previous sections.

It is worth noting what happens if the observation interval becomes as large as the interval to be predicted, that is $a \to (T-t)$. Gripenberg and Norros (1996) demonstrated that in this case $\rho_{H,a}$ tends to ρ_H or $\hat{\sigma}_{T,t,a}^2$ to $\hat{\sigma}_{T,t}^2$ respectively. So, concerning the variance, a limited historical observation interval is justified. Contrarily, the influence of additionally observed historical increments on the conditional mean will not vanish. However, recalling Fig. 5.1, this influence is strictly decreasing and can be roughly estimated. Hence, it is possible to focus on a limited historic interval when evaluating the information of the past. The results drawn from this finite observation will not differ too much from those that can be received taking an unlimited vantage point. Nevertheless, it turns out to be more comfortable to consider the theoretical case of unlimited information and we will focus thereon in the following.

5.3 A Conditional Fractional Itô Theorem

Having investigated the conditional properties of arithmetic fractional Brownian motion, we will now also consider the geometric price process S_t from a conditional point of view. For that purpose we introduce the notation of the conditional process $\hat{S}_s = S_s|[\omega_1]_t$. That is, we restrict the process to a part of the probability space $(\Omega, \mathfrak{A}, P)$, namely to the space generated by the equivalence class $[\omega_1]_t$, which is $([\omega_1]_t, \sigma([\omega_1]_t), \hat{P})$. The probability measure equals the conditional probability $\hat{P}$ so that for any process X the accordance of $\hat{E}(\hat{X}_T)$ and $E[X_T|\mathfrak{F}_t^H](\omega_1)$ immediately follows.

We further look at the dynamics of $\ln(\hat{S}_T)$ applying a conditional version of the fractional Itô theorem of (5.2):
For $s > t$, let $\hat{B}_s^H$ be the conditional process of fractional Brownian motion as above. For $F(s, \hat{B}_s^H)$ once continuously differentiable with respect to s and twice with respect to $\hat{B}_s^H$, we obtain under certain regularity conditions:

$$
\begin{aligned}
F(T, \hat{B}_T^H) = F(t, \hat{B}_t^H) &+ \int_t^T \frac{\partial}{\partial s} F(s, \hat{B}_s^H) \, ds \\
&+ \int_t^T \frac{\partial}{\partial x} F(s, \hat{B}_s^H) \hat{B}_s^H \, d\hat{B}_s^H \\
&+ \rho_H H \int_t^T (s-t)^{2H-1} \frac{\partial^2}{\partial x^2} F(s, \hat{B}_s^H) \left(\hat{B}_s^H \right)^2 ds.
\end{aligned}
\tag{5.15}
$$

As the process of geometric fractional Brownian motion S_t is a function of time t and fractional Brownian motion B_t^H, one can extend the relation of (5.15) replacing $F(t, \hat{B}_t^H)$ by $F(t, h(t, \hat{B}_t^H))$ with

$$
h(t, \hat{B}_t^H) := S_0 \exp \left(\mu t - \frac{1}{2} \sigma^2 t^{2H} + \hat{\sigma} B_t^H \right).
$$

One obtains a conditional fractional Itô theorem for functions of geometric fractional Brownian motion:

$$
\begin{aligned}
F(T, \hat{S}_T) = F(t, \hat{S}_t) &+ \int_t^T \frac{\partial}{\partial s} F(s, \hat{S}_s) \, ds \\
&+ \int_t^T \mu(s) \frac{\partial}{\partial x} F(s, \hat{S}_s) \hat{S}_s \, ds + \sigma \int_t^T \frac{\partial}{\partial x} F(s, \hat{S}_s) \hat{S}_s \, d\hat{B}_s^H \\
&+ \rho_H H \sigma^2 \int_t^T (s-t)^{2H-1} \frac{\partial^2}{\partial x^2} F(s, \hat{S}_s) \hat{S}_s^2 \, ds.
\end{aligned}
\tag{5.16}
$$

We sketch the derivation of (5.15) modifying the proof for the unconditional theorem of Bender (2003a). The idea is to show that the left hand side and the right-hand side have identical S-transforms. We recall some results related to the S-transform approach (for details, see Sect. 2.6). For $0 < H < 1$, the Riemann–Liouville fractional integrals were defined by

$$I_-^{H-\frac{1}{2}} f(x) = \frac{1}{\Gamma(H-\frac{1}{2})} \int_x^\infty f(s)(s-x)^{H-\frac{1}{2}}\, ds,$$

$$I_+^{H-\frac{1}{2}} f(x) = \frac{1}{\Gamma(H-\frac{1}{2})} \int_{-\infty}^x f(s)(x-s)^{H-\frac{1}{2}}\, ds,$$

while the operators $M_\pm^H$ were defined by

$$M_\pm^H f = K_H I_\pm^{H-\frac{1}{2}} f,$$

where
$$K_H = \Gamma\left(H + \frac{1}{2}\right)\sqrt{\frac{2H\Gamma(\frac{3}{2} - H)}{\Gamma(H+\frac{1}{2})\Gamma(2-2H)}}.$$

The S-Transform SF of the stochastic integral $F = \int_a^b X_t\, dB_t^H$ is the unique functional satisfying

$$SF(\eta) = S\left(\int_a^b X_t\, dB_t^H\right)(\eta) = \int_a^b S(X_t)(\eta)(M_+^H \eta)(t)\, dt := E^{Q_\eta}(F),$$

where Q_η is a measure that is defined by its Radon–Nikodym derivative with respect to the original measure P:

$$\frac{dQ_\eta}{dP} = \exp\left(\int_\mathbb{R} \eta(u)\, dB_u - \frac{1}{2}\int_\mathbb{R} \eta(u)^2\, du\right).$$

We also recall the fractional Girsanov formula: It states that if B_s^H is a fractional Brownian motion with respect to the measure P, then $\breve{B}_s^H$, defined via

$$\breve{B}_s^H = B_s^H - \int_0^s M_+^H \eta(u)\, du,$$

is a fractional Brownian motion with respect to Q_f.

Using this formula and applying the results concerning conditional moments of Sect. 5.2, we obtain the distribution of the $[\omega_1]_t$ - restricted process $\hat{B}_T^H$ with respect to Q_η to be normal with mean

$$\mu_{Q_\eta} = \breve{m}_{T,t} + \int_0^t M_+^H \eta(s)\, ds$$

and variance

$$\sigma_{Q_\eta}^2 = \rho_H (T-t)^{2H},$$

where

$$\check{m}_{T,t} = \check{B}_t^H + \int_{-\infty}^t g(T-t, s-t)\, d\check{B}_s^H(\omega_1)$$

is the conditional mean of the fractional Brownian motion $\check{B}_s^H$ and $\check{B}_s$ is the generating Brownian motion of $\check{B}_s^H$ as in Sect. 5.2.

We now readdress ourselves to the proof of (5.15). For this purpose, we investigate the S-transform of $F(s, \hat{B}_s^H)$. Modifying theorem 1.2.8 of Bender (2003a), we receive

$$S(F(s, \hat{B}_s^H)) = \int_{\mathbb{R}} F(s, u + \mu_{Q_\eta})k(\sigma_{Q_\eta}^2, u)\, du, \tag{5.17}$$

where $k(s,x) = \frac{1}{\sqrt{2\pi s}}\exp(\frac{-x^2}{2sx})$ is the so-called heat kernel. Differentiating both sides with respect to t, one obtains

$$\frac{d}{ds}S(F(s, \hat{B}_s^H))(\eta) = \int_{\mathbb{R}} \frac{\partial}{\partial s}F(s, u + \mu_{Q_\eta})k(\sigma_{Q_\eta}, u)\, du$$

$$+ M_+^H \eta(s) \int_{\mathbb{R}} \frac{\partial}{\partial x}F(s, u + \mu_{Q_\eta})k(\sigma_{Q_\eta}, u)\, du \tag{5.18}$$

$$+ \int_{\mathbb{R}} F(s, u + \mu_{Q_\eta})\frac{d}{ds}k(\sigma_{Q_\eta}, u)\, du.$$

Applying again theorem 1.2.8 of Bender (2003a), the first part on the right hand side equals $S\left(\frac{\partial}{\partial s}F(s, \hat{B}_s^H)\right)(\eta)$, while the second expression becomes $(M_+^H \eta)(s)S\left(\frac{\partial}{\partial x}F(s, \hat{B}_s^H)\right)(\eta)$. The third term can be rewritten using the fact that the function k satisfies the heat conduction equation $\frac{\partial}{\partial s}k = \frac{1}{2}\frac{\partial^2}{\partial x^2}k$. Using this and integrating by parts yields

$$\int_{\mathbb{R}} F(s,x)\frac{d}{ds}k(\sigma_{Q_\eta}, u)\, du = \left(\int_{\mathbb{R}} F(s,x)\frac{\partial^2}{\partial x^2}k(\sigma_{Q_\eta}, u)\, du\right)\frac{1}{2}\frac{d}{ds}(\sigma_{Q_\eta})$$

$$= \left(\left[F(s,x)\frac{\partial}{\partial x}k(\sigma_{Q_\eta}, u)\right]_{\mathbb{R}}\right.$$

$$\left. - \int_{\mathbb{R}} \frac{\partial}{\partial x}F(s,x)\frac{\partial}{\partial x}k(\sigma_{Q_\eta}, u)\right)\frac{1}{2}\frac{d}{ds}(\sigma_{Q_\eta})$$

$$= \left(\left[F(s,x)k(\sigma_{Q_\eta}, u)\left(-\frac{u}{s}\right)\right]_{\mathbb{R}}\right.$$

$$\left. - \left[\frac{\partial}{\partial x}F(s,x)k(\sigma_{Q_\eta}, u)\right]_{\mathbb{R}}\right.$$

$$\left. + \int_{\mathbb{R}} \frac{\partial^2}{\partial x^2}F(s,x)k(\sigma_{Q_\eta}, u)\right)\frac{1}{2}\frac{d}{ds}(\sigma_{Q_\eta}).$$

Note that by inserting $\pm\infty$, the first two terms equal zero. The last term can be reformulated using (5.17) and we obtain:

$$\int_{\mathbb{R}} F(s, u + \mu_{Q_\eta}) \frac{d}{ds} k(\sigma_{Q_\eta}, u)\, du = \frac{1}{2} \frac{d}{ds} (\sigma_{Q_\eta}) S\left(\frac{\partial^2}{\partial x^2} F(s, \hat{B}_s^H) \right)(\eta).$$

Hence, we can rewrite (5.18):

$$\frac{d}{ds} S(F(s, \hat{B}_s^H))(\eta) = S\left(\frac{\partial}{\partial s} F(s, \hat{B}_s^H) \right)(\eta) + (M_+^H \eta)(s) S\left(\frac{\partial}{\partial x} F(s, \hat{B}_s^H) \right)(\eta)$$

$$+ \frac{1}{2} \frac{d}{ds} (\sigma_{Q_\eta}) S\left(\frac{\partial^2}{\partial x^2} F(s, \hat{B}_s^H) \right)(\eta).$$

Integrating from t to T and rearranging the terms, we finally receive

$$S(F(T, \hat{B}_T^H))(\eta) = S(F(t, \hat{B}_T^H))(\eta) + S\left(\int_t^T \frac{\partial}{\partial s} F(s, \hat{B}_s^H)\, ds \right)(\eta)$$

$$+ S\left(\int_t^T \frac{\partial}{\partial x} F(s, \hat{B}_s^H)\, d\hat{B}_s^H \right)(\eta)$$

$$+ S\left(\rho_H H \int_t^T (s - t)^{2H-1} \frac{\partial^2}{\partial x^2} F(s, \hat{B}_s^H)\, ds \right)(\eta),$$

where we used the definition of the S-transform to replace the expression $\int_t^T (M_+^H \eta)(s) S\left(\frac{\partial}{\partial x} F(s, \hat{B}_s^H) \right)(\eta)\, ds$ by $S\left(\int_t^T \frac{\partial}{\partial x} F(s, \hat{B}_s^H)\, d\hat{B}_s^H \right)(\eta)$. Consequently, both sides of (5.15) have identical S-transforms and the result is proven.

In the following section, relation (5.16) will turn out to be very useful when pricing options in a fractional Brownian market.

5.4 Fractional European Option Prices

In this section, we are interested in the price at time t of a European call on S with maturity T and strike K.

As mentioned above, we assume the existence of a minimal period of time lying between two consecutive transactions. This assumption limits us with regard to the feasibility of pricing approaches based on no arbitrage arguments with a continuously adjusted replicating portfolio. Therefore, it seems to be natural to focus on preference based equilibrium pricing approaches. We do this in a very simple but all the more illustrative way, assuming risk-neutral investors, yet possessing and using information about the past. Hence, we consider the discounted conditional expected value of a contingent claim based on the observation of $[\omega_1]_t$:

$$C_{T,H}(t) = e^{-r(T-t)} E\left[\max(S_T - K)|\mathfrak{F}_t^H\right] \tag{5.19}$$

Typically, the pricing problem is solved defining a suitable measure under which expectations are taken. In the fractional Brownian market, we are faced with incompleteness. Probabilistic theory demonstrates that in this case the pricing measure is no longer unique. Unlike in the case of the complete Black–Scholes market, the choice of measure has to be motivated by arguments based on the assumed risk preferences of the investors. For example, Sottinen and Valkeila (2003) suggest using the measure Q^{SV} satisfying

$$E_{Q^{(SV)}}\left[e^{-r(T-t)}S_T\right] = S_t,$$

which they call the average risk neutral measure. It focuses on an equilibrium with respect to present time t and maturity T. Under this measure, the value of the stock in time T is on average unbiased. This measure has the advantage that it neither depends on the realized path nor on maturity T. On the other hand—as expectations are unconditional—, this equilibrium condition does not exploit the available information that can be taken from history.

Our pricing approach is the following: Concerning the equilibrium condition, we take on the idea of focusing on a two-time approach. However, we make two crucial modifications: Firstly, we account for the path-dependence of fractional Brownian motion by making the transition to an equilibrium condition based on conditional expectation. Additionally, we will not achieve the equilibrium by changing measure. Instead, we use the equilibrium condition to endogenously determine the unique constant drift rate of the underlying stock process. The implied logic of this is that in a world where all investors are risk-neutral, the basic asset cannot be of an arbitrary shape, but has to be in equilibrium itself.

For an observed history $\mathfrak{F}_t^H$ and a fix maturity T, we postulate that the discounted conditional expected value of the stock in time T equals its current value S_t:

$$E\left[e^{-r(T-t)}S_T|\mathfrak{F}_t^H\right] = S_t. \tag{5.20}$$

Note that this equation still holds under the physical measure. The equilibrium is brought about by an adjustment of the drift μ. There is a unique constant equilibrium drift rate which ensures (5.20) to hold. Note that—taking the vantage point of a risk-neutral investor—this equation indeed is in line with our fundamental understanding of equilibrium: The individuum should be indifferent between buying the stock and holding the amount S_t of the riskless asset. That is, under the physical measure, conditional expectation should equal the certainty equivalent, or more formally

$$E\left[S_T | \mathfrak{F}_t^H\right] = S_t e^{r(T-t)}, \tag{5.21}$$

which of course is the same as (5.20). Consequently, in a market that provides such an equilibrium, the basic pricing equation (5.19) is also to be interpreted with respect to the physical measure.

To exploit (5.21), we have to consider the conditional distribution of S_T given $[\omega_1]_t = \{\omega \in \Omega | B_s^H(\omega) = B_s^H(\omega_1), \forall s \in (-\infty, t]\}$. We further look at the conditional process $\hat{S}_t$ and apply the conditional version of the fractional Itô theorem (5.16) derived in the preceding section. With $F(s, \hat{S}_s) = \ln \hat{S}_s$ we have

$$\ln\left(\hat{S}_T\right) = \ln \hat{S}_t + \mu(T - t) - \frac{1}{2}\rho_H \sigma^2 (T - t)^{2H} + \sigma(\hat{B}_T^H - \hat{B}_t^H)$$

The first three terms being deterministic at time t, we obtain the distribution of $\ln\left(\hat{S}_T\right)$ by application of the results of Sect. 5.2 concerning the conditional moments of fractional Brownian motion (see eqs. (5.8)–(5.9)). We deduce that the logarithm of the conditional process $\hat{S}_T$ is normally distributed with the following moments:

$$m = \hat{E}\left(\ln\left(\hat{S}_T\right)\right) = E\left[\ln\left(\hat{S}_T\right) | \mathfrak{F}_t^H\right](\omega_1) \tag{5.22}$$

$$= \ln S_t + \mu(T - t) - \frac{1}{2}\rho_H \sigma^2 (T - t)^{2H} + \sigma \hat{\mu}_{T,t}$$

$$v = \hat{E}\left(\ln(\hat{S}_T) - m\right)^2 = E\left[(\ln\left(\hat{S}_T\right) - m)^2 | \mathfrak{F}_t^H\right](\omega_1) \tag{5.23}$$

$$= \rho_H \sigma^2 (T - t)^{2H}$$

where $\hat{\mu}_{T,t}$ and ρ_H are as in Sect. 5.2.

We further state that, $\ln(\hat{S}_T)$ being $N(m, v)$ distributed on $([\omega_1]_t, \sigma([\omega_1]_t), \hat{P})$, $\hat{S}_T$ must be log-normally distributed thereon with moments

$$M = \exp\left(m + \frac{1}{2}v\right) = S_t e^{\mu(T-t) + \sigma \hat{\mu}_{T,t}},$$

$$V = \exp(2m + 2v) - \exp(2m + v) = S_t^2 e^{2\mu(T-t)}\left(e^{\rho_H \sigma^2 (T-t)^{2H}} - 1\right).$$

But the mean M of the conditional process $\hat{S}_t$ equals the conditional mean of the process S_t. Hence we obtain

$$E\left[S_T | \mathfrak{F}_t^H\right] = M = S_t e^{\mu(T-t) + \sigma \hat{\mu}_{T,t}}. \tag{5.24}$$

Now we can exploit (5.21). Let

$$S_T = S_t + \int_t^T \bar{\mu} S_s \, ds + \int_t^T \sigma S_s \, dB_s^H$$

be the representation of the stock price process, where $\bar{\mu}$ is the adjusted drift rate. Using (5.24) and inserting into (5.21) we get

$$S_t e^{\bar{\mu}(T-t)+\sigma \hat{\mu}_{T,t}} = S_t e^{r(T-t)},$$

or

$$\bar{\mu}(T - t) = r(T - t) - \sigma \hat{\mu}_{T,t}. \tag{5.25}$$

The latter equation can be interpreted in the following way: The adjusted drift $\bar{\mu}(T - t)$ is split up into two parts. The first part equals the return one would receive from the riskless asset. In the Markovian case of classical Brownian motion that would already be all: the drift is shifted to equal the riskless return. However, in the case of a fractional Brownian market, there is an additional correction accounting for the evolution of the past. More precisely, we have an historically induced shift $-\sigma \hat{\mu}_{T,t}$ of the distribution. This means, a positive prediction for the random process of fractional Brownian motion results in a downward correction of the adjusted drift rate.

Note that this is not at all counter-intuitive. We explain the adjustment by three steps. Ex ante, investors have a first crude idea about the deterministic drift rate. In the first step of the adjustment, the investors extrapolate the positive evolution of the path into a positive adjustment of stock's future distribution. Next, they compare their predictions with the given current value of the stock and observe a mispricing. In our case, the discounted conditional expectation of the stock would exceed its current value. In the third step, investors react and update their expectations concerning the deterministic drift component which they obviously overestimated. In the given case, a downward correction of the deterministic drift results.

Summarizing we state the following relationship: the more promising the prediction of S_T due to the observation of the past, the more evident the mispricing of the (generally accepted) stock, and—for equilibrium reasons—the stronger the downward adjustment of the deterministic drift rate.

Combining relation (5.25) with (5.22) and (5.23) we obtain the conditional moments of $\ln(S_T)$ to be

$$m = \ln S_t + r(T - t) - \frac{1}{2}\rho_H \sigma^2 (T - t)^{2H}, \tag{5.26}$$

$$v = \rho_H \sigma^2 (T - t)^{2H}. \tag{5.27}$$

The associated density of the conditional process $\hat{S}_T$—which naturally is the conditional density of S_T based on the observation $[\omega_1]_t$—is as follows:

$$f(x)|_{[\omega_1]_t} = \frac{1}{x\sqrt{2\pi v}} e^{-\frac{1}{2}\frac{(lnx-m)^2}{v}} \, \mathrm{I}_{[x>0]}.$$

The well-known calculations lead to the following presentation for the price of the European call:

$$C_{T,H}(t) = e^{-r(T-t)} E_{Q^{(T,t)}} \left[\max(S_T - K)|\mathfrak{F}_t^H \right] \tag{5.28}$$

$$= S_t e^{m+\frac{1}{2}v - r(T-t)} N(d_1) - Ke^{-r(T-t)} N(d_2), \tag{5.29}$$

where

$$d_1^H = \frac{m + v - \ln K}{\sqrt{v}}, \tag{5.30}$$

$$d_2^H = \frac{m - \ln K}{\sqrt{v}} = d_1 - \sqrt{v}. \tag{5.31}$$

Inserting the terms for m and v of (5.26) and (5.27), we obtain the pricing formula for the fractional European call. The price of a fractional European call with strike K and maturity T valued by a risk-neutral investor is given by the following formula:

$$C_{T,H}(t) = S_t N(d_1^H) - Ke^{-r(T-t)} N(d_2^H), \tag{5.32}$$

where

$$d_1^H = \frac{\ln\left(\frac{S_t}{K}\right) + r(T-t) + \frac{1}{2}\rho_H \sigma^2 (T-t)^{2H}}{\sqrt{\rho_H}\sigma(T-t)^H},$$

$$d_2^H = \frac{\ln\left(\frac{S_0}{K}\right) + r(T-t) - \frac{1}{2}\rho_H \sigma^2 (T-t)^{2H}}{\sqrt{\rho_H}\sigma(T-t)^H} = d_1^H - \sqrt{\rho_H}\sigma(T-t)^H.$$

Following the same arguments as in the derivation of (5.32), we receive the price of the appropriate European put:

$$P_{T,H}(t) = Ke^{-r(T-t)} N(-d_2^H) - S_t N(-d_1^H). \tag{5.33}$$

For geometric Brownian motion, the put–call–parity pulls together the prices of options with the same underlying and contract parameters. Obviously, also in the fractional context, this fundamental relationship between call price, put price and stock price holds, as we have

$$C_{T,H}(t) - P_{T,H}(t) = S_t - Ke^{-r(T-t)}, \tag{5.34}$$

that is, we have a fractional put–call–parity. The proof is straightforward inserting the formulae (5.32) and (5.33) into (5.34).

Before investigating the obtained formulae more closely, we take a look at the special case of independent increments, as $H = \frac{1}{2}$. We know from Sect. 5.2 that in this case the information-based shift $\mu_{T,t}$ is zero and the variance reduction factor $\rho_{\frac{1}{2}}$ equals one. Hence, the conditional moments are

$$E_{Q^{(T,t)}} \left[B_t^{\frac{1}{2}} | \mathfrak{F}_t^{\frac{1}{2}} \right] = B_t,$$

$$Var_{Q^{(T,t)}} \left[B_T^{\frac{1}{2}} | \mathfrak{F}_t^H \right] = (T - t),$$

which are the well-known results of classical Brownian motion where the usage of historical information does not effectuate the distribution of the future.

Furthermore, we get for the equilibrium condition that the drift μ of the stock price process equals the riskless interest rate r. Using this, the conditional moments of the normally distributed log-price process are

$$m = \ln S_t + \left(r - \frac{1}{2}\sigma^2 \right)(T - t),$$

$$v = \sigma^2(T - t),$$

which again are identical to the unconditional moments. Inserting these special values into (5.29), (5.30) and (5.31), the classical Black–Scholes option pricing formula is received. On the other hand, the accordance can also be seen immediately when looking at the fractional pricing formulae and replacing ρ_H and H by 1 and $\frac{1}{2}$ respectively. Hence, the derived formulae are compatible extensions of the Black–Scholes option pricing formulae yielding the familiar result for the classical case.

We take a first look at the values of the fractional European options for different Hurst parameters H. Apparently, in the cases displayed in Fig. 5.4 and 5.5, an increasing Hurst parameter comes along with a decrease of the option value. One is tempted to argue that persistence reduces uncertainty whereas antipersistence increases the latter. For a short time to maturity $(T-t)$, the described interrelation is perfectly true. However, generally speaking, this is only half the truth. For larger values of $(T - t)$, a second effect comes into play which is based on the exponential shape of the unconditional variance function. A detailed examination of this phenomenon will be the subject of the following Sect. 5.5.

It is also possible to formally calculate partial derivatives of the option prices with respect to the parameters of the contract. These fractional Greeks should be interpreted with care. In our non-continuous setting, the partial derivatives

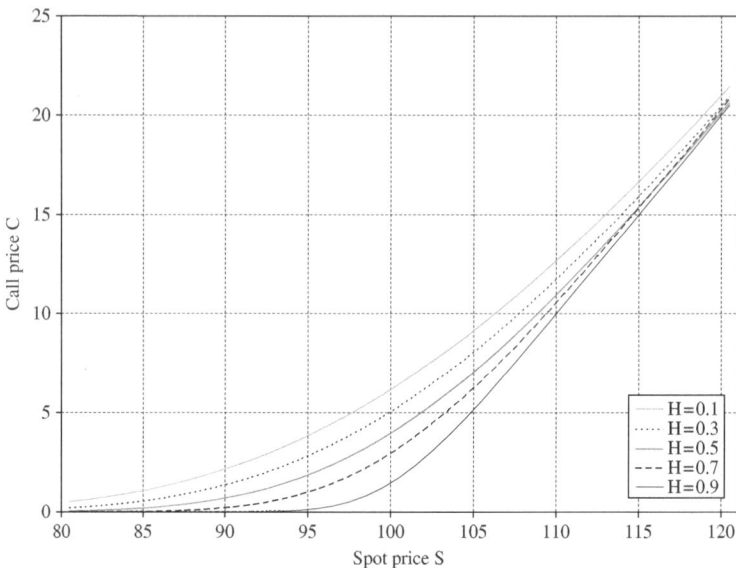

Fig. 5.4 Price of the fractional European call option with varying Hurst parameter H (chosen parameters: $r = 0.02$, $K = 100$, $\sigma = 0.2$, $T - t = 0.25$)

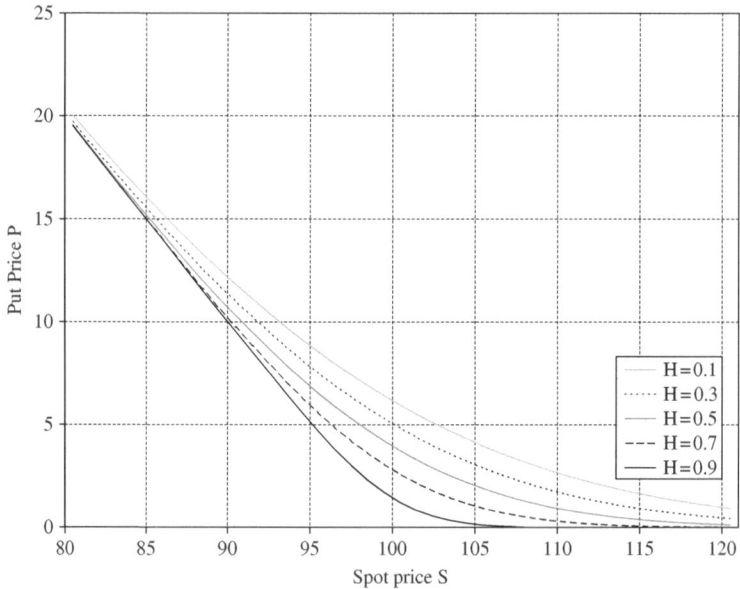

Fig. 5.5 Price of the fractional European put option with varying Hurst parameter H (chosen parameters: $r = 0.02$, $K = 100$, $\sigma = 0.2$, $T - t = 0.25$)

Table 5.1 The fractional Greeks

$\Delta_H = \frac{\partial C_H}{\partial S}$	$N(d_1^H)$	
$\Gamma_H = \frac{\partial^2 C_H}{\partial S^2}$	$\frac{\varphi(d_1^H)}{S_t\sqrt{\rho_H}\sigma(T-t)^H}$	
$\Theta_H = \frac{\partial C_H}{\partial T}$	$H\frac{S_t\varphi(d_1^H)\sqrt{\rho_H}\sigma}{(T-t)^{1-H}} + rKe^{-r(T-t)}N(d_2^H)$	
$\varrho_H = \frac{\partial C_H}{\partial r}$	$K(T-t)e^{-r(T-t)}N(d_2^H) - (T-t)S_tN(d_1^H)$	
$\Lambda_H = \frac{\partial C_H}{\partial \sigma}$	$S_t\varphi(d_1^H)\sqrt{\rho_H}(T-t)^H$	

can be considered as sensitivities concerning modifications of the initial parameters. However they are not suitable for hedging purposes. In particular,
a partial derivative with respect to current time t is not sensible. A dynamic
delta hedging as in the world of geometric Brownian motion is no longer
possible in our modified framework as it would imply arbitrage possibilities.
However, a sensitivity with respect to the date of expiration T can be derived
without any problems. Table 5.1 gives an overview of the partial derivatives
of the call price formula, the so-called fractional Greeks.

The proof of the formulae is straightforward. We stress the fact that as $H \to$
$\frac{1}{2}$, $(T-t)^H$ becomes $\sqrt{T-t}$, ρ_H tends to 1 and d_1^H becomes d_1 and the
well-known pa rameters of the Markovian case are obtained. So again, the
fractional solution in the limit also yields the results of classical Brownian
theory.

5.5 The Influence of the Hurst Parameter

The preceding results confirm the high degree of transferability of the classical
concepts into the fractional framework. However, an aspect of additional
interest arises from the consideration of the partial derivative with respect to
the Hurst parameter H, which will be denoted by η. To get an ex ante idea
of what we examine, recall that the Hurst parameter indicates the process-
immanent level of persistence. While $H = \frac{1}{2}$ ensures independent increments
and hence a Markovian process, values of H deviating from $H = \frac{1}{2}$ exhibit a
certain extent of dependence. The question is, in which manner the occurrence
of dependence influences the price of the fractional call.
We thus differentiate (5.32) with respect to H and get

$$\eta = \frac{\partial C}{\partial H} = S_t \varphi(d_1^H) \frac{\partial d_1^H}{\partial H} - K e^{-r(T-t)} \varphi(d_2^H) \frac{\partial d_2^H}{\partial H}$$

$$= S_t \varphi(d_1^H) \frac{\partial \left(\sqrt{\rho_H} \sigma (T-t)^H \right)}{\partial H} \tag{5.35}$$

$$= S_t \varphi(d_1^H) \frac{\partial \sqrt{v_T}}{\partial H},$$

where we introduced the notation $v_T = \sigma^2 \rho_H (T-t)^{2H}$. We first look at $\frac{\partial \rho_H}{\partial H}$ and differentiate the nominator $n(H) = \sin(\pi(H - \frac{1}{2}))\Gamma(\frac{3}{2} - H)^2$ and the denominator $d(H) = \pi(H - \frac{1}{2})\Gamma(2 - 2H)$ of ρ_H separately. For that purpose, note that $\Gamma'(x) = \Gamma(x)\psi_0(x)$ where ψ_0 denotes the digamma function. We get

$$\frac{\partial n}{\partial H} = \pi \cos\left(\pi \left(H - \frac{1}{2}\right)\right) \left(\Gamma \left(\frac{3}{2} - H\right)\right)^2$$

$$- \sin\left(\pi \left(H - \frac{1}{2}\right)\right) 2\Gamma\left(\frac{3}{2} - H\right) \Gamma\left(\frac{3}{2} - H\right) \psi_0\left(\frac{3}{2} - H\right)$$

$$= \left(\Gamma\left(\frac{3}{2} - H\right)\right)^2 \sin\left(\pi \left(H - \frac{1}{2}\right)\right)$$

$$\times \left[\pi \cot\left(\pi \left(H - \frac{1}{2}\right)\right) - 2\psi_0\left(\frac{3}{2} - H\right)\right],$$

$$\frac{\partial d}{\partial H} = \pi\Gamma(2 - 2H) - 2\pi \left(H - \frac{1}{2}\right) \Gamma(2 - 2H)\psi_0(2 - 2H)$$

$$= \pi\Gamma(2 - 2H) \left[1 - (2H - 1)\psi_0(2 - 2H)\right].$$

Using the quotient rule, we obtain

$$\frac{\partial \rho_H}{\partial H} = \rho_H \left[\pi \cot\left(\pi \left(H - \frac{1}{2}\right)\right) - 2\psi_0\left(\frac{3}{2} - H\right) - \frac{1}{H - \frac{1}{2}} + 2\psi_0(2 - 2H)\right].$$

We further make use of the following properties of the digamma function (see Abramowitz and Stegun (1972), Section 6.3):

$$\pi \cot(\pi x) = \psi_0(1 - x) - \psi_0(x),$$

$$\psi_0(x + 1) = \psi_0(x) + \frac{1}{x},$$

$$\psi_0(2x) = \frac{1}{2}\left(\psi_0(x) + \psi_0\left(x + \frac{1}{2}\right) + 2\ln 2\right).$$

Thus we can write

$$\frac{\partial \rho_H}{\partial H} = \rho_H \left(\psi_0 \left(\frac{3}{2} - H \right) - \psi_0 \left(H - \frac{1}{2} \right) - 2\psi_0 \left(\frac{3}{2} - H \right) \right.$$

$$\left. + \psi_0 (1 - H) + \psi_0 \left(\frac{3}{2} - H \right) + 2\ln 2 \right)$$

$$= \rho_H \left(\psi_0 (1 - H) - \psi_0 \left(H - \frac{1}{2} \right) - \frac{1}{H - \frac{1}{2}} + 2\ln 2 \right)$$

$$= \rho_H \left(\psi_0 (1 - H) - \psi_0 \left(H + \frac{1}{2} \right) + 2\ln 2 \right).$$

Putting these parts together, we can calculate the inner derivative $\frac{\partial v_T}{\partial H}$:

$$\frac{\partial v_T}{\partial H} = \frac{\partial \sigma^2 \rho_H (T - t)^{2H}}{\partial H}$$

$$= \sigma^2 \left(\rho_H \psi_0 (1 - H) - \psi_0 \left(H + \frac{1}{2} \right) \right.$$

$$\left. + 2\ln 2 (T - t)^{2H} + \rho_H 2\ln(T - t)(T - t)^{2H} \right)$$

$$= \rho_H \sigma^2 (T - t)^{2H} \left(\psi_0 (1 - H) - \psi_0 \left(H + \frac{1}{2} \right) + 2\ln 2 + 2\ln(T - t) \right)$$

$$= v_T \left(\psi_0 (1 - H) - \psi_0 \left(H + \frac{1}{2} \right) + 2\ln 2 + 2\ln(T - t) \right). \qquad (5.36)$$

Note that the digamma function $\psi_0(x)$ for $x > 0$ is strictly monotonic increasing but concave, the negative axis of ordinates being vertical asymptote as x tends to zero (see Fig. 5.6). Therefore, the difference $\psi_0(1 - H) - \psi_0(H + \frac{1}{2})$ is strictly monotonic decreasing for $0 < H < 1$, and its maximum is received for $H \to 0$. For parameters $H < \frac{1}{4}$, the difference is positive whereas for larger values of H it becomes negative. In this case we get

$$\lim_{H \to 0} \left[\psi_0 \left(H + \frac{1}{2} \right) - \psi_0 (1 - H) \right] = \psi_0(1) - \psi_0 \left(\frac{1}{2} \right)$$

$$= 2\ln 2 + \gamma - \gamma = 2\ln 2,$$

where γ denotes the Euler–Mascheroni constant.

Summarizing, we can state the following results, denoting by τ the time to maturity, that is $\tau = T - t$: The partial derivative of the fractional call price C with respect to the Hurst parameter H is given by

$$\eta = S_t \varphi(d_1^H) \frac{1}{2\sqrt{v_T}} v_T \left(\psi_0 (1 - H) - \psi_0 \left(H + \frac{1}{2} \right) + 2\ln 2 + 2\ln(T - t) \right)$$

$$= S_t \varphi(d_1^H) \sqrt{\rho_H} \sigma (T - t)^H \frac{(\psi_0(1 - H) - \psi_0(H + \frac{1}{2}) + 2\ln 2 + 2\ln(T - t))}{2}.$$

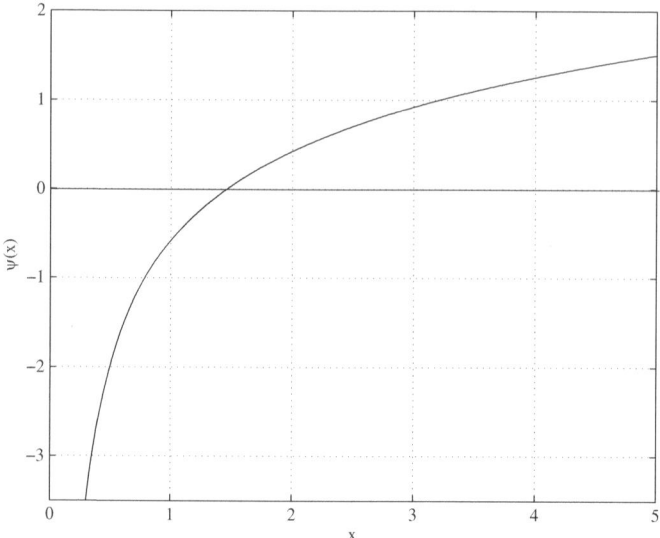

Fig. 5.6 The digamma function on $\mathbb{R}_+$

Concerning the sign of this sensitivity, one can easily show that we have two qualitatively different cases:

- For a fix $\tau \leq \frac{1}{4}$, it holds:

$$\frac{\partial C}{\partial H}(H) < 0, \quad \forall\, 0 < H < 1.$$

- For a fix $\tau > \frac{1}{4}$, there exists a critical Hurst parameter $0 < \bar{H} < 1$, so that:

$$\frac{\partial C}{\partial H}(\bar{H}) = 0,$$
$$\frac{\partial C}{\partial H}(H) > 0, \quad \forall\, 0 < H < \bar{H},$$
$$\frac{\partial C}{\partial H}(H) < 0, \quad \forall\, \bar{H} < H < 1.$$

The second case can be further specified:

- For $\frac{1}{4} < \tau < 1$, it follows that $\bar{H}$ and consequently the maximum of the call value lie in the antipersistent area.
- For $\tau = 1$, the critical Hurst parameter is $\bar{H} = 0.5$. So the case of serial independence yields the highest call price.
- For $\tau > 1$, the parameter $\bar{H}$ and therefore the maximal call value lie in the persistent area.

The results are immediate consequences of the preceding observations as well as of the properties of the natural logarithm. For example, one can prove the strict negativity of η for $\tau < \frac{1}{4}$ as follows:

$$\eta = S_t \varphi(d_1^H) \sqrt{\rho_H} \sigma (T-t)^H \frac{(\psi_0(1-H) - \psi_0(H + \frac{1}{2}) + 2\ln 2 + 2\ln(T-t))}{2}$$

$$< S_t \varphi(d_1^H) \sqrt{\rho_H} \sigma (T-t)^H \frac{(2\ln 2 + 2\ln 2 + 2\ln(\frac{1}{4}))}{2} = 0.$$

Figs. 5.7–5.9 illustrate these characteristics graphically, showing the relation between the Hurst parameter H and the call price for a fix initial price S_t.

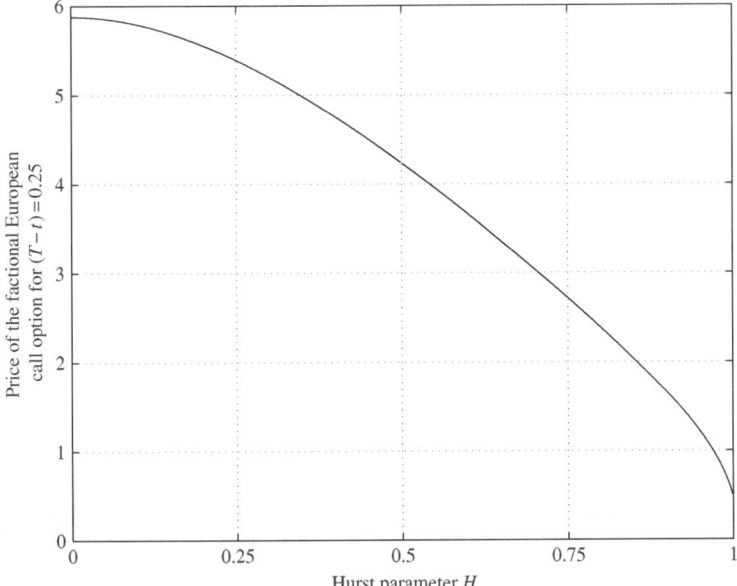

Fig. 5.7 Maturity effect on the relation between the price of the fractional European call and Hurst parameter H for maturity $T = 0.25$ (chosen parameters: $r = 0.02$, $S = 100$, $K = 100$, $\sigma = 0.2$)

In order to be able to explain this phenomenon, we recall that according to (5.36) the main effect arises from the product $\rho_H \sigma^2 \tau^{2H}$, which is the variance v_T of the normal distribution of the conditional logarithmic stock price. But, with increasing H, the factors of v_T generate converse effects.

Tracking the factor ρ_H over the range of possible parameters, it starts for $H \to 0$ with 0.5, then increases as antipersistence decreases, takes it maximum 1 for serial independence as $H = 0.5$ and—with an increasing level of persistence—finally decreases again (see Fig. 5.2). Besides the case $H = \frac{1}{2}$, the factor ρ_H is smaller than one, hence ρ_H concentrates the distribution,

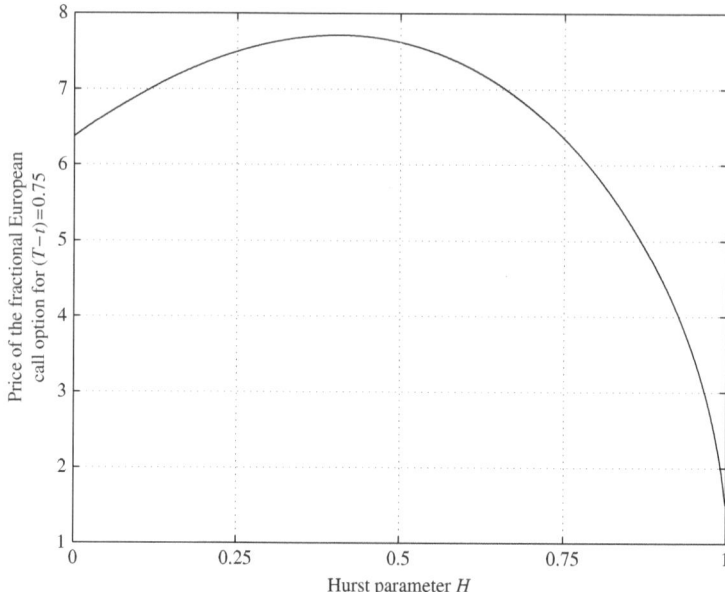

Fig. 5.8 Maturity effect on the relation between the price of the fractional European call and Hurst parameter H for maturity $T = 0.75$ (chosen parameters: $r = 0.02$, $S = 100$, $K = 100$, $\sigma = 0.2$)

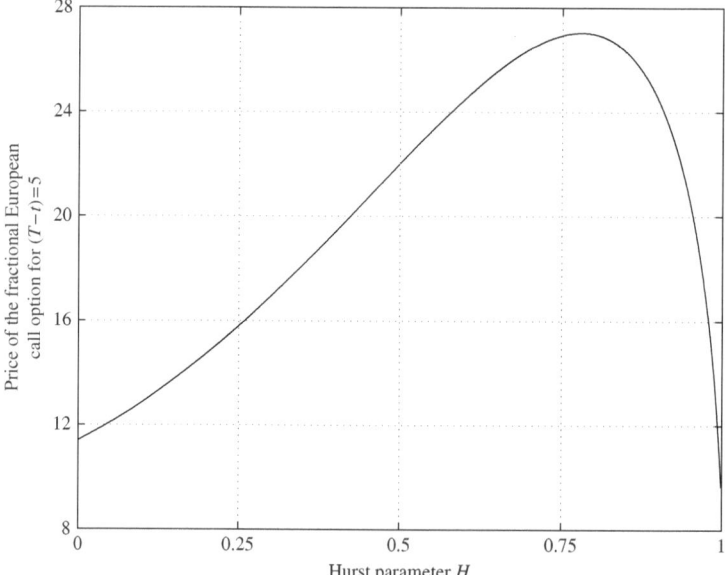

Fig. 5.9 Maturity effect on the relation between the price of the fractional European call and Hurst parameter H for maturity $T = 5$ (chosen parameters: $r = 0.02$, $S = 100$, $K = 100$, $\sigma = 0.2$)

what we from now on call the narrowing effect.

On the other hand, the behaviour of the term τ^{2H} depends on the value of τ (see Fig. 5.3). A higher Hurst parameter implies a higher exponent. For $\tau > 1$, this higher exponent enlarges the unconditional variance τ^{2H} whereas for $\tau < 1$—meaning short time horizons—small Hurst parameters increase the variance. This is in line with our basic understanding of fractional Brownian motion: The higher the level of persistence, the smoother the paths—which means little variance in the short run—and the larger are the cyclical long-time deviations from the mean. Increasing antipersistence however implies rougher paths fluctuating more closely around the mean which yields an increasing variance on short horizons and a decreasing variance on long horizons. The particular effect of τ^{2H} is an unconditional effect. It is not related to any information about the past and is also valid for the unconditional variance. It is further referred to as the power effect.

Thus, the resulting effect also depends on the scale of τ. We first consider nearby distributional forecasts ($\tau < 1$). In the persistent parameter domain, the total effect is clear: Both effects reduce the variance as H increases. Concerning antipersistence we observe that for τ smaller than $\frac{1}{4}$, the power effect totally dominates the narrowing effect, yielding a strictly decreasing variance (see Fig. 5.7). For $\frac{1}{4} < \tau < 1$ however, there is always a range of antipersistent parameters $0 < H < \bar{H} < \frac{1}{2}$, where the positive narrowing effect dominates the negative power effect. For antipersistent parameters larger than $\bar{H}$, the power effect is the outweighting one (see Fig. 5.8).

If we consider larger time horizons with $\tau > 1$, it is the antipersistent domain that yields uniqueness of the total effect: Both the narrowing effect and the power effect enlarge the variance as H increases between zero and one half. For the case of persistence, starting from $H = 0.5$, first the power effect is the controlling one, but only up to a critical value $\bar{H}$. From here on, the high degree of persistence and thereby predictability outbalances the increase of unconditional variance and reduces the conditional variance (see Fig. 5.9).

Summarizing, we can state that the relation between the Hurst parameter H and the price of a European call generally exhibits a hump-shaped curve. The larger the maturity τ, the larger is the critical value $\bar{H}$. For very small maturities $\tau < \frac{1}{4}$, $\bar{H}$ equals zero and the hump degenerates to a decreasing line.

A brief look at the limit of the call price as H tends to 1 stresses the above observations and confirms our intuition with regard to fractional Brownian motion. With an increasing Hurst parameter, we obtain an increasing level of dependence: that is, the future price of the underlying becomes less volatile or uncertain. In the limit we distinguish between two cases. For $S > e^{-r(T-t)}K$, d_1^H and d_2^H tend to infinity and for the call price we actually receive the difference between the initial stock price and the discounted strike price. On the other hand, if we have $S < e^{-r(T-t)}K$, d_1^H and d_2^H tend to $-\infty$, and the call price tends to zero. So in the case of perfect dependence, either the contract value is zero right from the beginning or we get a simple forward contract under certainty.

5.6 The Influence of Maturity and the Term Structure of Volatility

The preceding section showed that time to maturity τ matters concerning the effects of serial correlation. In particular, the relation between Hurst parameter H and call price C was qualitatively mainly influenced by this time to expiration. The next proximate step is to investigate the characteristics of the call price when depicted over τ. Figures 5.10–5.12 accomplish this, the upper picture gives the antipersistent case, the lower one stands for persistence. The picture in the middle is the case of the classical Brownian motion model where increments are serially uncorrelated. In all three cases, the three qualitatively different types of moneyness are specified. Note that for the case $H = 0.5$, this kind of picture was already provided by Cox and Rubinstein (1985).

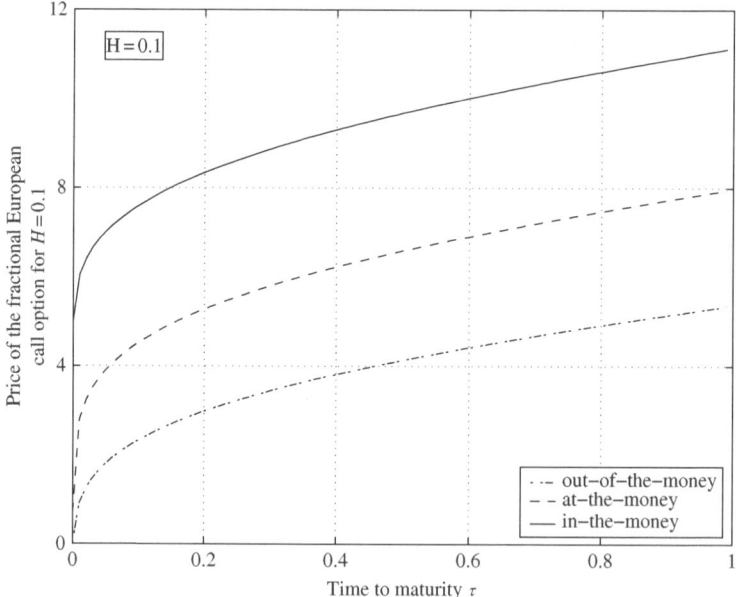

Fig. 5.10 Value of the European call option over time to maturity for $H = 0.1$ and different types of moneyness (chosen parameters: $r = 0.02$, $S = 100$, $\sigma = 0.2$)

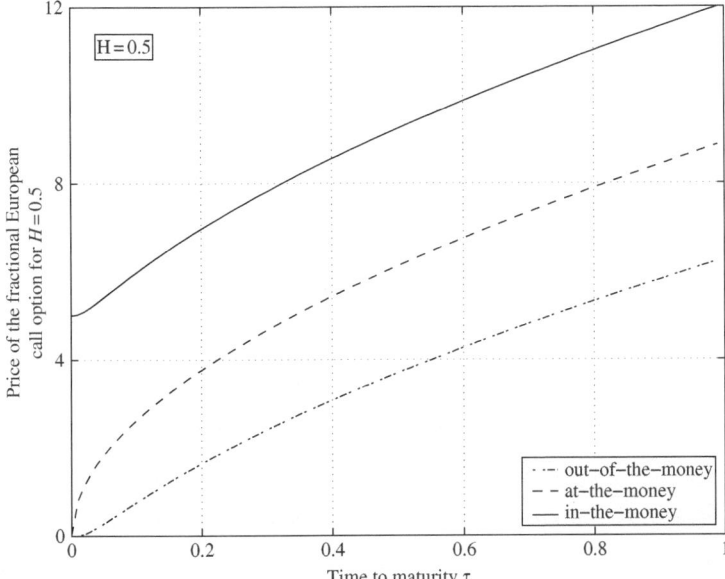

Fig. 5.11 Value of the European call option over time to maturity for $H = 0.5$ and different types of moneyness (chosen parameters: $r = 0.02$, $S = 100$, $\sigma = 0.2$)

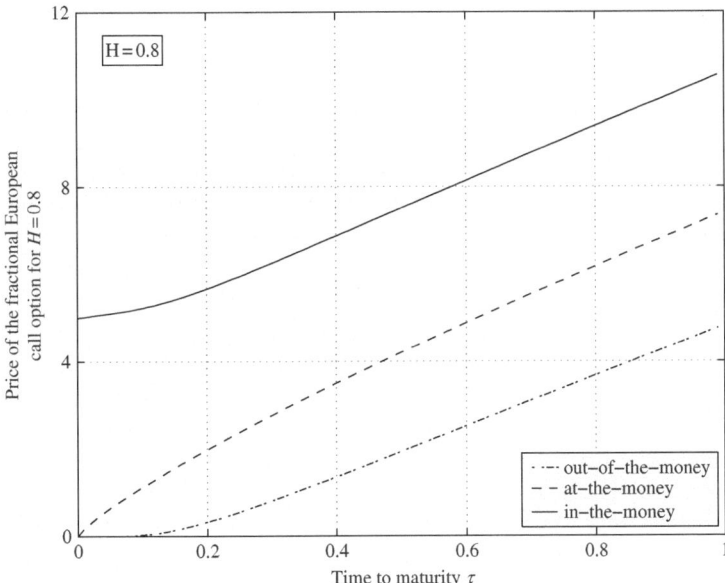

Fig. 5.12 Value of the European call option over time to maturity for $H = 0.8$ and different types of moneyness (chosen parameters: $r = 0.02$, $S = 100$, $\sigma = 0.2$)

The two effects by which the Hurst parameter affects the call price were the narrowing effect and the power effect. While the first one is driven by the constant conditionality factor ρ_H, the latter one exerts its influence via the unconditional part of the variance $(T - t)^{2H}$. It is therefore time-dependent and can be rediscovered in Fig. 5.12: We now compare the pictures of the persistent and antipersistent case respectively with the one discussed by Cox and Rubinstein (1985) taking the independent case $H = 0.5$ as a starting point. In this case, the curve for the in-the-money call is concavely shaped, for the at-the-money and out-of-the-money calls, we observe slightly S-shaped curves. For $H < \frac{1}{2}$, the relation between time to maturity and the variance now is no longer linear but a concave one. Consequently—as the call price is driven by the extent of uncertainty—the three lines in the Figs. 5.10–5.12 become more concave as we introduce antipersistence. For the same reason, we get a more convex character (in the sense of less concavity or a stronger S-shape respectively) for the case of persistence as the latter implies a convexly shaped relation between time to maturity and uncertainty of the contract. Comparing for example only the curves of the options being at-the-money, we consequently observe the following: With decreasing time to maturity, the loss of the option value is most striking in the case of antipersistence, whereas in the case of persistence it is less pronounced than in the classical case.

Figure 5.13 displays the three-dimensional relation between Hurst parameter, time to maturity and the value of a at-the-money call. A sectional view being parallel to the maturity axis produces pictures of the type of the Figs. 5.10–5.12, whereas a sectional view parallel to the axis of Hurst parameters would be of the kind of the Figs. 5.7–5.9.

The preceding pictures give only a crude and imprecise impression of the fact that something new comes into play. Though the curves of the call value over maturity change, there seems to be no fundamental innovation when curves become more concave or convex. However, it is exactly the behavior with respect to the dimension of time that exhibits new desirable properties. This will become clear when looking at the so-called term structure of volatility which depicts the implied Black–Scholes volatilities over time to maturity.

One of the properties of the classical option pricing models by Black and Scholes (1973) and Merton (1973), that is often criticized, is the fact, that real market prices of derivatives do not—as stated by these models—imply constant volatility over time and moneyness. Instead, when taking real prices as given and calculating backwards the model-implied volatilities, one obtains so called volatility smiles or volatility smirks (see e.g. Derman and Kani (1994) or Dupire (1994)). Many alternative models have been proposed that are able to account for this phenomenon like models with time-dependent or stochastic volatility. The usual testing procedure is to calculate the implied Black–Scholes volatilities of option prices derived by the new model and check whether the model is able to yield smiles and smirks similar to those

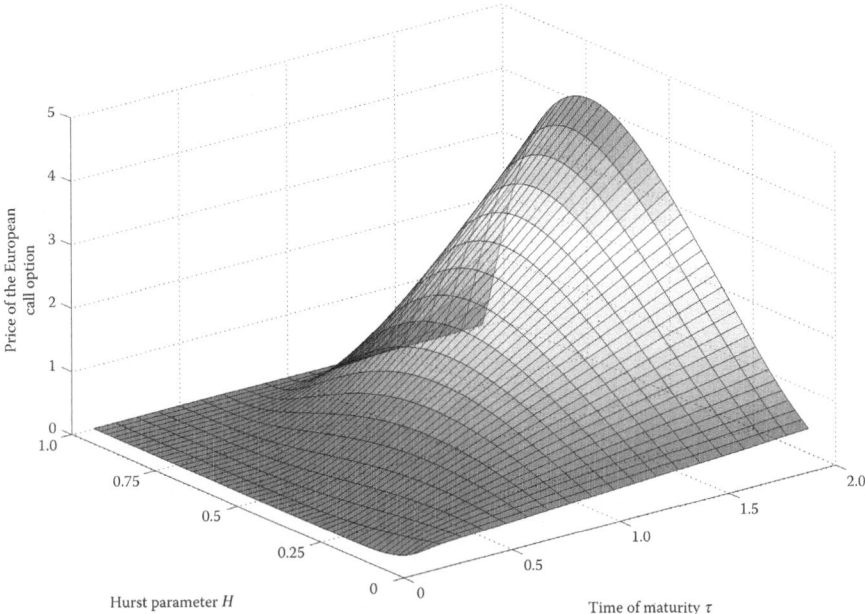

Fig. 5.13 Value of the European call option over time to maturity and Hurst parameter H (chosen parameters: $r = 0.02$, $S = 100$, $K = 100$, $\sigma = 0.2$)

observed for real market data. However, most of the models that are able to generate the desired volatility curves over time to maturity need numeric methods when calculating the implied Black–Scholes volatility.

As an advantage of our model, numerical techniques are not necessary in our case. Instead, the implied volatility can be derived analytically. The fractional model only replaces the classical variance $\sigma(T - t)$ by the conditional fractional variance $\sigma \rho_H (T - t)^{2H}$. So, the implied Black–Scholes volatility $\tilde{\sigma}$ has to satisfy the following equation:

$$\tilde{\sigma}(T - t) = \sigma \rho_H (T - t)^{2H},$$

which leads to

$$\tilde{\sigma} = \sigma \rho_H (T - t)^{2H-1}.$$

Obviously, there is no dependence of the implied volatility on moneyness. Hence, one obtains only flat lines with respect to this dimension. Actually, the shape of the implied volatility over time to maturity is much more interesting. The relation is fully described by the exponent $(2H - 1)$ yielding root functions for the persistent case and hyperbolas for Hurst parameters smaller than one half. Figure 5.14 depicts the implied term structure of volatility for different Hurst parameters.

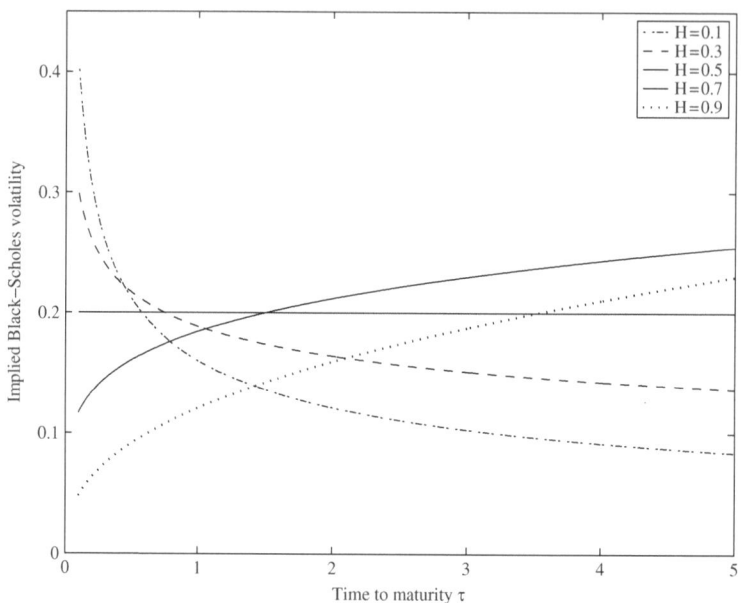

Fig. 5.14 Implied Black–Scholes volatility for different Hurst parameters over time to maturity (chosen parameters: $r = 0.02$, $S = 100$, $K = 100$, $\sigma = 0.2$)

The shape of the curves agree with our previous considerations. In the case of antipersistence, fractional Brownian motion induces a high level of uncertainty in the very short run, where its paths heavily fluctuate. In the long run however, the mean-reverting character leads to less uncertainty than classical Brownian motion. On the other hand, if we address ourselves to persistence, we have smoother paths and consequently less uncertainty in the short run, whereas the long-range dependent character reinforces deviations from the mean and implies an increase of uncertainty compared to the classical case. With regard to empirical relevance, Hurst parameters $H > \frac{1}{2}$ are usually found to be more adequate (see e.g. Lo and MacKinley (1988) or Willinger et al. (1999)). We stress that the concave term structure of volatility that in this case results from our model, can be refound in a different framework of current interest: the option pricing model by Carr and Wu (2003) who assume stocks to be driven by a finite moment log-stable process.

Chapter 6
Risk Preference-Based Option Pricing in the Fractional Binomial Setting

In this chapter, we take on the setting presented in Chap. 3. There, we showed that binomial trees can be used to approximate the process of geometric fractional Brownian motion. The framework is free of arbitrage if investors are restricted to trade only on certain nodes.

Concerning the valuation of derivatives, the classical binomial approach applies the idea of backward calculation. Starting from the terminal nodes, it is possible to go backwards step by step, relating two posterior nodes to one antecedent and using no arbitrage arguments. Within our modified setting, we are faced with the obvious problem that a step-by-step procedure is no longer possible as we could not adopt absence of arbitrage on this minimal time scale. On the other hand, if we leave the minimal time scale and only look at intervals where we can ensure absence of arbitrage, the way backwards is not unique: If the arbitrage-free interval is divided into n steps, we have to relate 2^n subsequent nodes to only one prior node.

As we will assert risk-neutrality of the market participants, one could have the idea of simply taking expectations. From this vantage point, assets should be valued by their conditional mean. However, the inconsiderate usage of this kind of valuation ignores the fact that one of the two basic assets—the risky one—is not a martingale. Hence, its discounted conditional expected value in time t generally does not equal its current value. Per se, this does not cause any problem as long as predictability cannot be exploited. For our basic setting with two assets, this was ensured by our restricted framework which we introduced in Chap. 3. However, if one coonsiders a further asset being related to the stock, it is not reasonable to apply a pricing rule that does not hold for the basic risky asset. Doing so, one obtains a disequilibrium between derivative and underlying. We give a short example to see that such a partial disequilibrium leads to an arbitrage possibility.

S. Rostek, *Option Pricing in Fractional Brownian Markets.*
Lecture Notes in Economics and Mathematical Systems.
© Springer-Verlag Berlin Heidelberg 2009

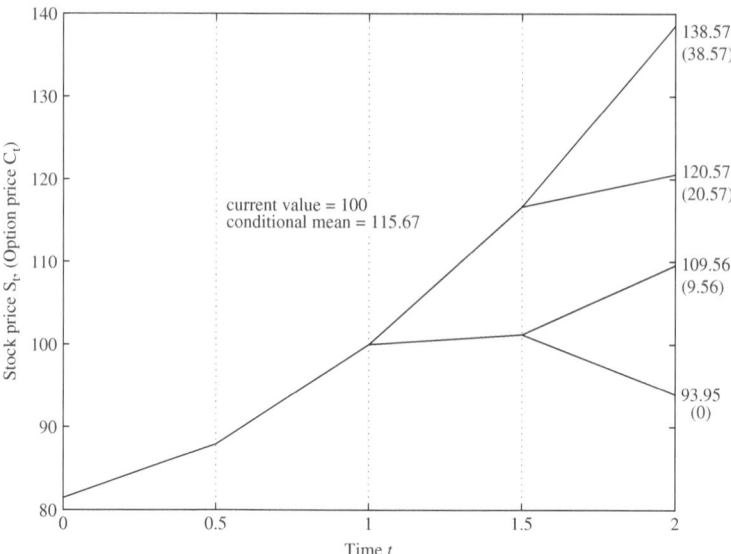

Fig. 6.1 Evolution of a persistent geometric Brownian motion (underlying stock) and terminal payoffs of a call option for a specific historic path under the physical measure (chosen values: $H = 0.8$, $S_0 = 100$, $\mu = 0$, $\sigma = 0.2$, $K = 100$, $r = 0$)

Figure 6.1 shows the evolution of a stock following a geometric fractional Brownian motion with Hurst parameter $H = 0.8$. For reasons of simplicity we choose the riskless interest rate to be zero. The history up to time $t = 1$ is known and the future distribution is influenced by this history. The physical measure is the one under which each upwards or downwards step occurs with the probability of one half. Evidently, under this physical measure, the conditional mean of the terminal stock value does not equal its current value. We look at a European call option with maturity in $T = 2$ and strike $K = 100$. In Fig. 6.1, the respective terminal payoffs are given in brackets. If we priced the option by its conditional expected terminal payoff, we would get the option price

$$\hat{C}_{1,2}^{0.8} = E\left[C_2|\mathfrak{F}_1\right] = \frac{1}{4}\left(38.57 + 20.57 + 9.56 + 0\right) = 17.18.$$

It is however easy to show, that this valuation leaves space for an arbitrage opportunity. Consider the following portfolio R of an investor held from time $t = 1$ until $T = 2$:

- a short position of one unit of the call option, initial value $\hat{C}_{1,2}^{0.8} = 17.18$,
- a long position of one unit of the stock, initial value $1 * S_1 = 100$,
- a short position of the riskless bond with total value of 90 monetary units.

Consequently, buying the portfolio R in time $t = 1$ yields a net cash flow of $17.18 - 1 * 100 + 90 = 7.18$, which is positive. On the other hand, in time $T = 2$, we have four possible states of nature which we number serially and denote in brackets. The particular payoffs of the liquidation of the portfolio R then are:

$$R(1) = -38.57 + 138.57 - 90 = 10,$$
$$R(2) = -20.57 + 120.57 - 90 = 10,$$
$$R(3) = -9.56 + 109.56 - 90 = 10,$$
$$R(4) = 0 + 93.95 - 90 = 3.95.$$

In all four cases, the portfolio yields a non-negative value. Hence, an arbitrage opportunity is given if this kind of pricing procedure is applied.

Obviously, both the idea of backward calculation by building replicating portfolios for each node and the mere calculation of conditional expectation are inappropriate methods for pricing purposes in the fractional binomial context. Instead, we will now concentrate on equilibrium pricing. In principle, there are two possible ways to solve the partial disequilibrium. The following sections will deal with these two possibilities, a total equilibrium approach on the one hand and a relative equilibrium approach on the other hand. In the limit, both approaches will yield the same result, however, the respective motivations will be quite different.

The rest of the chapter is organized as follows: We will start with an analogue to the two-time pricing approach that we introduced for the continuous time case in the preceding chapter. In the next section, we will consider an alternative way of two-time valuation using a change of measure. We proceed by extending our equilibrium to all tradable points in time introducing so-called multi-time equilibria. In particular, we will investigate two possibilities how this can be done. Throughout the chapter, we will accompany all of our considerations with some examples.

6.1 The Two-Time Total Equilibrium Approach

The first possibility how to solve the problem of disequilibrium mentioned above, is to force the market to be in a total equilibrium. In this case, the process of the underlying asset is assumed to have a certain design which ensures it to be in equilibrium itself. The implied logic is the following: In a world where in time t all investors are risk-neutral, the basic risky asset cannot have an arbitrary drift μ, but there is only one possible constant drift. The latter is path-dependent and equilibrates the stock price over the period of interest. Therefore, it is quasi endogenous. With this drift, the discounted conditional expectation of the stock value in T equals its current value in time t. The risk-neutral investors then price all derivatives that are based on the stock also by their discounted conditional mean.

To be more precise, from current time t until maturity T, the process S has to have the unique constant drift rate μ_{adj} under which it satisfies

$$E\left[S_T|\mathcal{F}_t\right] = S_t \left(1 + r/n\right)^{(T-t)n}. \tag{6.1}$$

Note that this is exactly the same idea we applied in the preceding chapter for the continuous time setting. Due to the complexity of the Wick product, the endogenous equilibrium drift can in general only be calculated numerically. Furthermore, it depends on the number of approximation steps.

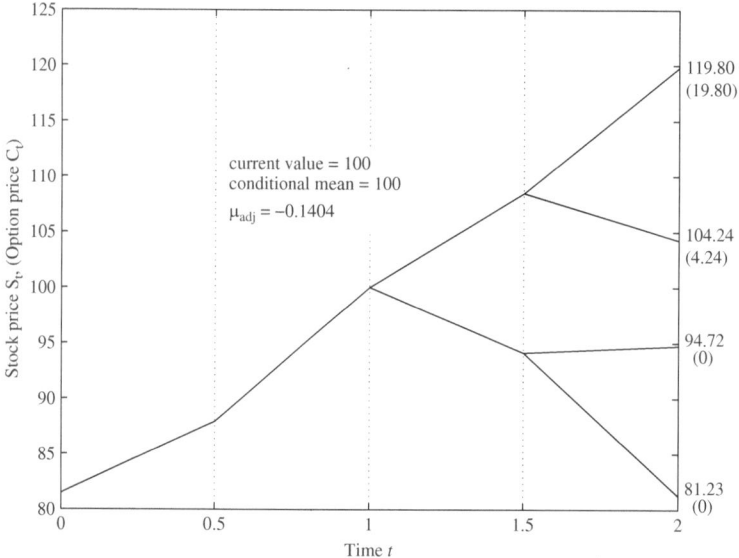

Fig. 6.2 Evolution of a persistent geometric Brownian motion (underlying stock) and terminal payoffs of a call option for a specific history after adjusting a path-dependent drift component (chosen values: $H = 0.8$, $S_0 = 100$, $\sigma = 0.2$, $K = 100$, $r = 0$)

Figure 6.2 exemplifies the described procedure graphically: For a given historical path up to time $t = 1$, one chooses a constant drift component so that the remaining random distribution satisfies $E\left[S_2|\mathcal{F}_1\right] = S_1$. The value of the adjusted drift rate μ_{adj} as well as the resulting stock prices are given in the picture. As the measure remains unchanged, the calculation of option prices is quite easy. In the example depicted in Fig. 6.2, we obtain for a European call option valued in time $t = 1$ and maturing in $T = 2$ the discounted conditional mean:

$$C_{tot}^{0.8}(1,2) = E\left[C_2|\mathfrak{F}_1\right] = \frac{1}{4}\left(19.8044 + 4.2442 + 0 + 0\right) = 6.0122.$$

It can be easily shown that this price does not allow for an arbitrage. We now take a look at what happens if one refines the partition of the time intervals. From the theoretical point of view, the result is clear: As seen in Chap. 3, with an increasing number of steps, the binomial distribution of the discrete arithmetic Brownian motion converges to the normal distribution of its continuous time counterpart. Accordingly, the discrete geometric binomial process exhibits a lognormal limit distribution. By the adjustment of the drift, the discounted conditional mean equals the current value whereas the conditional variance remains unchanged. Consequently, the price of the call option in the discrete time model should tend to the result that we obtained for continuous time in Sect. 5.4. For different levels of fineness and varying Hurst parameters, the option values can be seen in Table 6.1. Concerning the persistent parameter domain, the respective price of the call seems to converge to the value of the continuous case. The latter can be calculated by putting the respective parameters into formula (5.32) of Sect. 5.4. For the antipersistent values of H one would also presume convergence, however, the prices seem to be overestimated. The speed and preciseness of the convergence mainly depends on the quality of the approximation concerning the conditional variance which we investigated in Sect. 3.3. Ibidem, the problem occurred that in our finite-memory approach we only allow for a limited history whereas the continuous time framework assumes an infinite history. This difference matters most for Hurst parameters standing for slight antipersistence, that is, in a range from 0.3 to 0.4 (see Sect. 3.3).

Note that the choice of an adjustment by a constant drift parameter over the

Table 6.1 Prices of a European call option for varying Hurst parameters H and different numbers n of approximating steps per unit of time, (chosen parameters: $t = 1$, $T = 2$, $S_t = 100$, $\sigma = 0.2$, $r = 0$, $K = 100$)

$C_{tot}^H(1,2)$	$n = 2$	$n = 4$	$n = 6$	$n = 8$	$n = \infty$
$H = 0.2$	7.8890	7.8045	7.7094	7.6453	7.4363
$H = 0.4$	7.7713	7.9841	8.0147	8.0149	7.8892
$H = 0.6$	7.3646	7.6327	7.7301	7.7809	7.8629
$H = 0.8$	6.0122	6.5521	6.5708	6.6281	6.6663

whole interval (t, T) can be viewed as focusing on only the two most important points in time t and T. This is analogous to the equilibrium condition we introduced in the continuous time framework and the reason why we named it two-time approach. The in-between values do not display equilibrium values.

More precisely we do not have $E[S_s|\mathcal{F}_1] = S_1$ for points in time $1 < s < 2$. We will readdress ourselves to this problem later on.

6.2 The Two-Time Relative Equilibrium Approach

The idea of an endogenously-adjusted drift that compensates for the historical trend caused by persistence might seem artificial. However, it is nothing but the consequence of postulating that any tradable asset has to be in equilibrium itself. Meanwhile, there is another possibility left to equilibrate the market: We still assume risk-neutral investors and introduce an equilibrium condition taking the vantage point of a relative pricing approach. One regards the price of the basic asset or underlying as given without wondering for a moment how it came about. All other assets that are related to the underlying have to be in a relative equilibrium and their prices are derived from the accepted basic price. The information of this given stock value is used to update the probability measure and so the problem of asset pricing turns into the calculation of a conditional mean. More precisely, we will no longer use the physical measure, but an equivalent one, that takes into account the difference between current value of the underlying and its conditional mean. To summarize this outline, assets are valued by their expected value of the terminal payoff—conditioned on all available information about the past and put into a relative equilibrium to the given value of the underlying.

For any pair of current time t and maturity T, we define the conditional average risk neutral measure $Q_{t,T}$ to be the measure satisfying

$$E_{Q_{t,T}}[S_T|\mathcal{F}_t] = S_t (1 + r/n)^{(T-t)n}. \qquad (6.2)$$

This measure is not necessarily unique. It is known from financial theory that absence of arbitrage ensures existence of an equivalent measure. Furthermore, uniqueness follows market completeness (see e.g. Musiela and Rutkowski (2005), p. 63–75). In lack of completeness we have to impose further conditions in order to be able to identify a unique measure. Recall that we introduced the price process as a discrete version of a geometric fractional Brownian motion with constant drift μ. It is natural to postulate that under the new measure $Q_{t,T}$, the stock also follows a geometric fractional Brownian motion. This requirement is expressed by the attribute 'equivalent'. However, there is at least still one open question which is the specification of the drift under the new measure. One could either restrain it to be constant, or, instead, allow for a time-varying drift, being deterministic or even stochastic. In the present case, the equivalent measure is unique if and only if we confine ourselves to a constant drift. In all other cases, the number of degrees of freedom exceeds the number of equilibrium conditions which we reduced to one by focusing on a two-time valuation.

In order to derive the new measure, we first calculate the equilibrating drift $\bar{\mu}$. The calculation is the same as in the section of the total equilibrium and in general it can only be done using numerical methods. The difference to the procedure before is that this time the drift parameter is interpreted to display the deterministic behavior of the process under the equivalent measure, not of the physical one. Moreover, looking at the dynamics of the process S_s from t to T, we obtain under the new measure $Q_{t,T}$:

$$S_j = S_{j-1} \diamond \left(1 + \bar{\mu}/n + \sigma/n \left(B_j^{H(n)} - B_{j-1}^{H(n)}\right)\right) \tag{6.3}$$

Consequently, for this time interval, the logarithm of the stock price under $Q_{t,T}$ follows the arithmetic process

$$\ln S_j = \ln S_{j-1} + \bar{\mu}/n + \sigma/n \left(B_j^{H(n)} - B_{j-1}^{H(n)}\right) \tag{6.4}$$

We now want to determine the measure $Q_{t,T}$ explicitly. In particular, we are looking for transition probabilities which ensure that the stock price process satisfies the dynamics of (6.3) and (6.4). For this purpose we recall the construction of the binomial process. In Sect. 3.2, we introduced the binomial approximation of an arithmetic fractional Brownian motion without drift. We used independent binomial random variables ξ_i taking the values ± 1 with probability $\frac{1}{2}$ each. The n-th approximation of the fractional Brownian motion in time t, that divided each unit time interval into n steps, was then given by equation (3.7):

$$B_s^{H(n)} := \int_0^s z^{(n)}(s,u)dW_u^{(n)} = \sum_{i=1}^{[ns]} n \int_{\frac{i-1}{n}}^{\frac{i}{n}} z\left(\frac{[ns]}{n}, v\right) dv \frac{1}{\sqrt{n}} \xi_i^{(n)}.$$

For the interval $[t, T]$, we now change the probabilities of the realizations of the ξ_i. Instead of $\frac{1}{2}$ each, we define for $nt < i \le nT$ the i-th binomial random variable ξ_i to take the value $+1$ with probability

$$p_i = \frac{1}{2}\left(1 + \frac{\bar{\mu} - \frac{1}{2}\sigma^2(T-t)^{2H-1}}{n\sqrt{n}\int_{\frac{i-1}{n}}^{\frac{i}{n}} z\left(\frac{[nT]}{n}, u\right) du}\right), \tag{6.5}$$

and the value -1 with probability

$$1 - p_i = \frac{1}{2}\left(1 - \frac{\bar{\mu} - \frac{1}{2}\sigma^2(T-t)^{2H-1}}{n\sqrt{n}\int_{\frac{i-1}{n}}^{\frac{i}{n}} z\left(\frac{[nT]}{n}, u\right) du}\right). \tag{6.6}$$

Under this measure, for $nt < i \leq nT$ the random variables ξ_i have the (uncentered) moments

$$E\left[\xi_i\right] = \frac{\bar{\mu} - \frac{1}{2}\sigma^2(T-t)^{2H-1}}{n\sqrt{n}\int_{\frac{i-1}{n}}^{\frac{i}{n}} z\left(\frac{[nT]}{n}, v\right) dv},$$

$$E\left[\xi_i^2\right] = 1.$$

Note that the ξ_i are no longer identically distributed but still independent. Accordingly, if the history up to time t is known, we get for the n-th approximation $B_T^{H(n)}$ of B_T^H the conditional mean

$$E\left[B_T^{H(n)}|\mathcal{F}_t\right] = B_t^{H(n)} + \sum_{i=nt+1}^{[nT]} n\int_{\frac{i-1}{n}}^{\frac{i}{n}} z\left(\frac{[nT]}{n}, v\right) du \frac{1}{\sqrt{n}} \frac{\bar{\mu} - \frac{1}{2}\sigma^2(T-t)^{2H-1}}{n\sqrt{n}\int_{\frac{i-1}{n}}^{\frac{i}{n}} z\left(\frac{[nT]}{n}, v\right) dv}$$

$$= B_t^{H(n)} + \sum_{i=nt+1}^{[nT]} \frac{\bar{\mu} - \frac{1}{2}\sigma^2(T-t)^{2H-1}}{n}$$

$$= B_t^{H(n)} + \bar{\mu}(T-t) - \frac{1}{2}\sigma^2(T-t)^{2H}$$

and the conditional second moment

$$E\left[\left(B_{T,(n)}^H|\mathcal{F}_t\right)^2\right] = \sum_{i=nt+1}^{[nT]} \left(n\int_{\frac{i-1}{n}}^{\frac{i}{n}} z\left(\frac{[nT]}{n}, u\right) du \frac{1}{\sqrt{n}}\right)^2 \xrightarrow{n\to\infty} \rho_H(T-t)^{2H}.$$

The proof of the second moment relation is the same as in the case without drift (see Sect. 3.2), because the second moments of the ξ_i did not change. Summarizing, with the adjusted probabilities, the binomial process above approximates an arithmetic fractional Brownian motion with drift $\bar{\mu} - \frac{1}{2}\sigma^2 t^{2H-1}$. Consequently, it describes the desired process of equation (6.4). The respective geometric process has drift $\bar{\mu}$ and therefore fulfils equation (6.3).

Obviously, relations (6.5) and (6.6) uniquely define the measure $Q_{t,T}$, under which the original tree yields the constant equilibrium drift $\bar{\mu}$ and condition (6.2) is satisfied. Although we have in our example two different probabilities p_1 and p_2, there is still only one degree of freedom which is the equilibrating drift $\hat{\mu}$ and which determines the transition probabilities. Furthermore, absence of arbitrage ensures that the derived transition probabilities lie between zero and one. In Fig. 6.3, the original tree of fractional Brownian motion is endowed with these equilibrating probabilities which we calculated numerically.

Note that, in our example, the adjustment of the transition probabilities exhibits the direction we would have expected. As the original distribution under the physical measure in Fig. 6.1 was heavily biased upwards, the

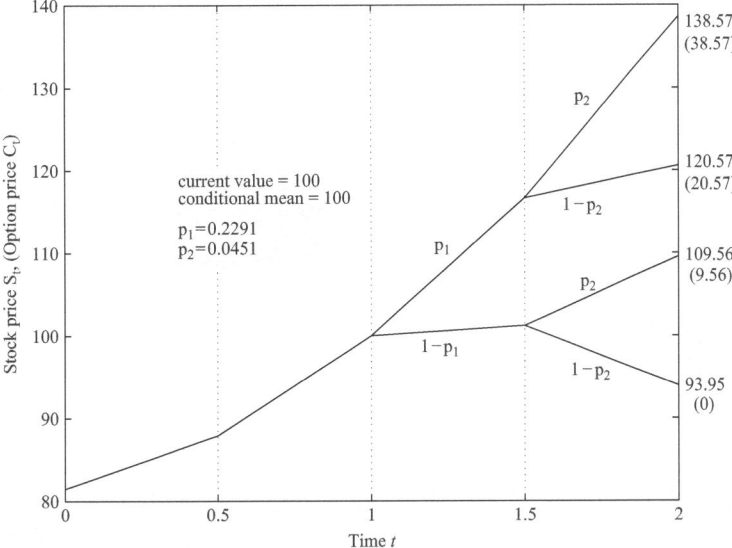

Fig. 6.3 Evolution of a persistent geometric Brownian motion (underlying stock) and terminal payoffs of a call option for a specific historic path with adjusted measure $Q_{t,T}$ (chosen values: $H = 0.8$, $S_0 = 100$, $\mu = 0$, $\sigma = 0.2$, $K = 70$, $r = 0$)

equivalent measure has to account for this bias and puts less weight on the upwards steps. Under the new measure, asset prices can be derived by calculating their discounted conditional expected payoff. The European call option with strike $K = 100$ and maturity $T = 2$ then has the following value in $t = 1$:

$$
\begin{aligned}
C_{rel}^{0.8}(1,2) = E_{Q_{1,2}}\left[C_2|\mathfrak{F}_1\right] = \ & 0.2299 * 0.0451 * 38.5718 \\
& + 0.2299 * 0.9549 * 20.5741 \\
& + 0.7701 * 0.0451 * 9.5635 \\
& + 0.7701 * 0.9549 * 0 \\
= \ & 5.2475.
\end{aligned}
$$

Having a look at the respective value of the total equilibrium approach $C_{rel}^{0.8}(1,2)$ in the preceding section, the difference is obvious. The partial equilibrium approach seems to be much farer from the limit value than that calculated above. However, if we make the partition over time more fine, the method of adjusting the measure also yields a good approximation. The call prices also seem to converge—at least for the parameters standing for persistence—to the value of the continuous time case (see Table 6.2).

The valuation by simply taking expectations is possible as—like in the classical Brownian case—the change of measure adjusts the drift. In the case of

Table 6.2 Prices of a European call option for varying Hurst parameters H and different numbers n of approximating steps per unit of time, (chosen parameters: $t = 1$, $T = 2$, $S_t = 100$, $\sigma = 0.2$, $r = 0$, $K = 100$)

$C_{rel}^{H}(1,2)$	$n = 2$	$n = 4$	$n = 6$	$n = 8$	$n = \infty$
$H = 0.2$	7.9041	7.8038	7.7447	7.6513	7.4363
$H = 0.4$	7.8113	7.9457	7.9793	7.9994	7.8892
$H = 0.6$	7.3930	7.6473	7.7299	7.7388	7.8629
$H = 0.8$	5.2475	6.5126	6.5477	6.5826	6.6663

fractional Brownian motion, we have two different types of drift components: the deterministic drift component and a path-dependent drift component which is due to the peculiarity of the fractional market and results from the memory of the process. By introducing the equivalent measure, the distribution is shifted in a way so that the stock price has the equilibrium drift rate $\bar{\mu}$. The sum of this deterministic drift $\bar{\mu}$ and of the path-dependent drift component equals the riskless interest rate.

When comparing this with the procedure known from classical Brownian theory, one notes a strong similarity. There, the price process is a semimartingale and the equivalent measure removes the drift and makes it equal to the riskless rate. In our setting, this is not sufficient as one has to account for the path-dependence of fractional Brownian motion which implies one additional component. The 'target drift rate' is therefore not r but the presented term $\bar{\mu}$ which includes both the riskless interest rate and the trend evolving from the historical path.

From the conceptual point of view, things are one-to-one comparable. However, there are technical differences: The prediction part (or path-dependent drift) in the fractional framework depends on both the available information (or current time t) and the prediction horizon (that is, maturity T). Consequently, the equilibrium drift $\bar{\mu}$ and likewise the equilibrium measure have to be redetermined for each new pair t and T. This is a crucial difference to the classical theory, where the equivalent measure does not depend on current time t or maturity T.

Two further problems stem from the same fact: On the one hand, a measure that satisfies the equilibrium condition (6.2) for a pair of points in time (t, T), will in general not fulfill the respective condition for other points in time (s, S). To put it differently, even under the new measure, the process is not a martingale. On the other hand, if one introduces additional degrees of freedom by allowing for a non-constant drift, the measure $Q_{t,T}$ is no longer uniquely determined by means of (6.2). We will show that these two problems are closely related to each other.

6.3 Multi-Time Equilibrium Approaches

While in the preceding sections we stressed the difference between the total
and the relative equilibrium approach, we now state the following: Albeit the
approaches differ heavily concerning their motivation and interpretation, the
calculus is more or less identical. In both cases, the equilibrium condition is
used to derive a certain drift component that equilibrates the process. In the
total equilibrium this is done with respect to the physical measure, whereas
in the relative equilibrium the drift adjustment is caught by the transition
to the equivalent measure. The mere calculation is however the same, and as
seen above, in the limit, both approaches lead to the same result. We will now
focus on problems that do not depend on the choice of equilibrium stated. For
sake of simplicity, we will forbear from writing out both cases in full. Unless
stated differently, the following verbalizations hold for both vantage points.
Whenever it is necessary to denote formulae, we will choose the notation of
the total equilibrium approach.

Evidently, for the two-time equilibrium approaches, the in-between stock values S_s for $t < s < T$ do not represent equilibrium values: We postulated
the equilibrium condition to hold only for the pair of points in time t and
T. As long as we focus on European options and additionally abstain from
in-between tradability, this will cause no problem. Though, for example a
sensible valuation of American options should not be possible under these
circumstances.

If we weaken the assumption that the equilibrating drift has to be constant,
we are provided with additional possibilities to impose equilibrium conditions
for in-between points in time. We will first investigate the possibilities arising
from a deterministic specification of the drift and then proceed and allow for
a path-dependent equilibrium drift.

6.3.1 Multi-Time Equilibria with Respect to Current Time t

If we allow the adjusted drift to be deterministic, the number of degrees of
freedom equals the number of steps $N_1 = (T - t)n$ lying between t and T,
where n is the number of steps per unit of time. Besides the fundamental
equilibrium condition (6.2), it is hence possible, to satisfy $N_1 - 1$ further
equilibrium conditions. In principle, for all discrete $t < s < T$, one could
therefore add the equilibrium conditions

$$E\left[S_s|\mathcal{F}_t\right] = S_s \left(1 + r/n\right)^{(s-t)n}. \tag{6.7}$$

However, it suffices to require condition (6.7) to hold for all nodes on the tradable grid of the binomial tree. Note that the tradable grid was defined by the very nodes between which the time interval is sufficiently large so that arbitrage possibilities cannot occur. The nodes lying on not admissible time scales need not fulfill (6.7). Consequently, on each minimal tradable time interval again, we can choose the drift to be piecewise constant. If we have k steps between two tradable nodes, the total number of equilibrium conditions is $N_2 = N_1/k$. Hence, we also get N_2 different drift adjustments.

Figure 6.4 displays the described procedure for the case $t = 1$, $T = 2$ and $r = 0$. The first picture shows the process before adjusting the equilibrating drift rates. The chosen number of in-between steps is $N_1 = 4$, the number k that ensures absence of arbitrage is 2, which yields $N_2 = 4/2 = 2$. The resulting equilibrium conditions are

$$E\left[S_{1,5}|\mathcal{F}_1\right] = S_1 \tag{6.8}$$

$$E\left[S_2|\mathcal{F}_1\right] = S_1. \tag{6.9}$$

The deterministic endogenous drift rates are denoted by $\mu_{adj}^{1-1.5}$ for the time interval $[1, 1.5]$ and $\mu_{adj}^{1.5-2}$ for the time interval $[1.5, 2]$. They are uniquely determined by the equilibrium conditions (6.8) and (6.9). The respective values are given in the second picture of Fig. 6.4, where the adjusted binomial tree is depicted.

Note that, according to equation (6.9), the basic equilibrium condition of the two-time equilibrium approach is still satisfied. As long as a deterministic drift rate is chosen, the transition from constancy to time-variability of the endogenous drift does not change the distribution of the terminal values. All terminal nodes are shifted by the same amount. The main innovation of the multi-time approach is the way how the adjustment is proportioned over the interval $[t, T]$. Consequently, the valuation of a European option should not change vitally and the comparison of Table 6.3 with the Tables 6.1 and 6.2 shows that this is indeed true.

By means of the additional equilibrium conditions, it is now possible to price American options. These options can be exercised at all states of nature and points in time lying on the tradable grid. In the example of Fig. 6.4, there is one possibility of early exercise at $t^* = 1.5$. Like the valuation of American options in the classical context, we now decide backwards whether the investor should hold or carry out the option. That is, the investor compares the payoff of an immediate exercise with the expected payoff in case he keeps the option. Again, we take the example of a call option with strike price $K = 100$. For each of the four nodes at time $t^* = 1.5$ in the lower picture of Fig. 6.4 (denoted by the superscripts uu, ud, du, dd respectively), we can easily calculate the value of an instantaneous exercise

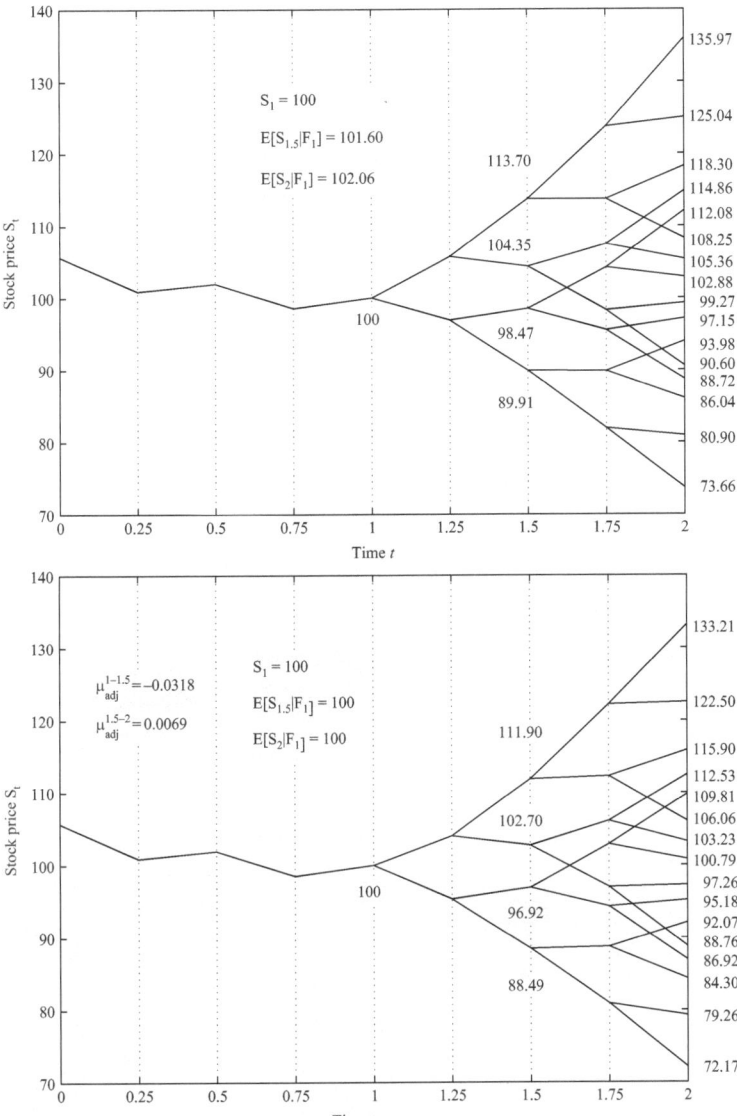

Fig. 6.4 Evolution of a persistent geometric fractional Brownian motion (underlying stock) and terminal payoffs of a call option for a specific historic path before and after adjusting a deterministic drift rate (chosen parameters: $H = 0.8$, $S_1 = 100$, $r = 0$, $\sigma = 0.2$)

Table 6.3 Prices of a European call option for varying Hurst parameters H and different numbers n of approximating steps per unit of time, derived by a multi-time equilibrium with respect to current time t (chosen parameters: $t = 1$, $T = 2$, $S_t = 100$, $\sigma = 0.2$, $r = 0$, $K = 100$, $k = 2$)

$C^H_{mult}(1,2)$	$n = 2$	$n = 4$	$n = 6$	$n = 8$	$n = \infty$
$H = 0.2$	7.8890	7.8045	7.7270	7.6453	7.4363
$H = 0.4$	7.7713	7.9841	8.0270	8.0149	7.8892
$H = 0.6$	7.3646	7.6327	7.7251	7.7767	7.8629
$H = 0.8$	6.5469	6.5339	6.5708	6.5693	6.6663

$$C^{uu}_{ex}(1.5) = \max[111.90 - 100, 0] = 11.90,$$
$$C^{ud}_{ex}(1.5) = \max[102.70 - 100, 0] = 2.70,$$
$$C^{du}_{ex}(1.5) = \max[96.92 - 100, 0] = 0,$$
$$C^{dd}_{ex}(1.5) = \max[88.49 - 100, 0] = 0.$$

The according values of holding the option are calculated by the conditional means of the terminal option payoffs:

$$C^{uu}_{hold}(1.5) = E\left[\max[S(2) - 100, 0]|\mathcal{F}^{uu}_{1,5}\right] = \frac{1}{4}(33.21 + 22.50 + 15.90 + 6.06)$$
$$= 19.42,$$

$$C^{ud}_{hold}(1.5) = E\left[\max[S(2) - 100, 0]|\mathcal{F}^{uu}_{1,5}\right] = \frac{1}{4}(12.53 + 3.23 + 0 + 0)$$
$$= 3.94,$$

$$C^{du}_{hold}(1.5) = E\left[\max[S(2) - 100, 0]|\mathcal{F}^{uu}_{1,5}\right] = \frac{1}{4}(9.81 + 0.79 + 0 + 0)$$
$$= 2.65,$$

$$C^{dd}_{hold}(1.5) = E\left[\max[S(2) - 100, 0]|\mathcal{F}^{uu}_{1,5}\right] = \frac{1}{4}(0 + 0 + 0 + 0)$$
$$= 0.$$

Obviously, in all four states of nature, the option right of early exercise is not made use of. Consequently, we get the values of the American option at time $t^* = 1.5$:

$$C^{uu}_{am}(1.5) = 19.42,$$
$$C^{ud}_{am}(1.5) = 3.94,$$
$$C^{du}_{am}(1.5) = 2.65,$$
$$C^{dd}_{am}(1.5) = 0.$$

Going one further step backwards, we compare for the one possible state of nature in time $t = 1$ the value of an early exercise

$$C_{ex}(1) = \max[100 - 100, 0] = 0,$$

which is zero, as the option is at-the-money, whereas the value of holding the option is

$$C_{hold}(1) = E\left[C_{am}(1.5)|\mathcal{F}_1\right] = \frac{1}{4}\left(19.42 + 3.94 + 2.65 + 0\right) = 6.50.$$

Therefore, we obtain

$$C_{am}(1) = 6.50.$$

This value is quite close to that of the European option, which comes as no surprise: In our example, the right of premature exercise does not generate an additional value, as it is always favorable to keep the option.

Summarizing we state: If we introduce the possibility of a deterministic time-varying drift, the stock price can be put into equilibrium at all tradable points in time, related to the current stock value S_t. However, this does not imply that when relating two in-between points in time to each other, the respective stock values are in equilibrium. More precisely, by lack of the Markov property of the process, this will indeed not be the case.

6.3.2 Local Multi-Time Equilibria

Even more opportunities arise if we allow for a path-dependent equilibrium drift. It is then possible to impose local equilibrium conditions: We stipulate that for each pair of subsequent points in time the respective stock prices have to be in equilibrium. The condition needs to hold for all possible (and admissible) states of nature, that is, all nodes of the tradable grid. For the example in the upper picture of Fig. 6.4, we would hence introduce the following equilibrium conditions

$$
\begin{aligned}
E\left[S_{1,5}|\mathcal{F}_1\right] &= S_1, \\
E\left[S_2|\mathcal{F}_{1,5}^{uu}\right] &= S_{1.5}^{uu}, \\
E\left[S_2|\mathcal{F}_{1,5}^{ud}\right] &= S_{1.5}^{ud}, \\
E\left[S_2|\mathcal{F}_{1,5}^{du}\right] &= S_{1.5}^{du}, \\
E\left[S_2|\mathcal{F}_{1,5}^{dd}\right] &= S_{1.5}^{dd},
\end{aligned}
\tag{6.10}
$$

where the superscripts describe the historical path of the process, e.g. ud standing for an upward movement followed by a downward step. By these conditions, the binomial tree is divided into five non-overlapping partial trees. On each partial tree, it is therefore possible to determine an endogenous drift rate so that the respective equilibrium condition is fulfilled. Within the respective part of the tree, the drift rate is chosen to be constant, implying uniqueness of the adjustment. With $N_2 + 1 = N_1/k + 1$ tradable points in time, we obtain the number of tradable nodes to be the sum of $N_2 + 1$ terms

$$N_3 = 2^0 + 2^k + 2^{2k} + \ldots + 2^{N_2 k} = \frac{1 - 2^{N_1 + k}}{1 - 2^k},$$

and the number of equilibrium conditions is

$$N_4 = 2^0 + 2^k + 2^{2k} + \ldots + 2^{(N_2 - 1)k} = N_3 - 2^{N_2 k} = \frac{1 - 2^{N_1}}{1 - 2^k}.$$

In the example above, we had $N_1 = 4$ steps, $k = 2$ as the fineness of the tradable grid, $N_2 + 1 = 3$ tradable points in time, $N_3 = 21$ tradable nodes and $N_4 = 5$ equilibrium conditions which were given by the system of equations (6.10).

One special case is worth mentioning: If $k = 1$, and all nodes of the binomial grid are tradable, the drift rate varies from each node to the next one. Most interestingly, in this case, the resulting drift-adjusted binomial tree is even recombining.

Although the idea of a path-dependent drift component might be appealing at first glance, there is at least one evident drawback of this concept. The introduction of a stochastic or path-dependent drift rate, changes the second moment of the distribution of the process. To see why and how, we fix the tradable grid to be the points in time $t = t_0, t_1, t_2, \ldots, t_{N_2} = T$ and discretize the in-between intervals by further non-tradable nodes. Then, conditioned on time t_{j-1}, the logarithm of the stock price at the next tradable node $\ln\left(S_{t_j}\right)$ tends to a normally-distributed random variable with conditional moments

$$m_j = E\left[\ln\left(S_{t_j}\right) | \mathcal{F}_{t_{j-1}}\right] = \ln S_{t_{j-1}} + r(t_j - t_{j-1}) - \frac{1}{2}\rho_H \sigma^2 (t_j - t_{j-1})^{2H},$$

$$v_j = E\left[\left(\ln\left(S_{t_j}\right) - E\left[\ln\left(S_{t_j}\right) | \mathcal{F}_{t_{j-1}}\right]\right)^2 | \mathcal{F}_{t_{j-1}}\right] = \rho_H \sigma^2 (t_j - t_{j-1})^{2H}.$$

This results from the properties of the binomial approximation (see Chap. 3) and the conditional fractional Itô theorem of Sect. 5.3.

We now assume the tradable grid to be equidistant, that is, for each $j = 1, \ldots, N_2$ we have $t_j - t_{j-1} = \frac{T-t}{N_2}$. Applying the law of iterated expectation, we obtain the distribution of $\ln\left(S_T\right)$ conditioned on the information available at time t:

$$
\begin{aligned}
m_{T,t} &= E\left[\ln\left(S_T\right)|\mathcal{F}_t\right] \\
&= E\left[E\left[\ldots E\left[E\left[\ln\left(S_T\right)|\mathcal{F}_{t_{N_2-1}}\right]|\mathcal{F}_{t_{N_2-2}}\right]|\ldots\right]|\mathcal{F}_t\right] \\
&= \ln S_t + \sum_{j=1}^{N_2} r(t_j - t_{j-1}) - \frac{1}{2}\rho_H\sigma^2\left(t_j - t_{j-1}\right)^{2H} \\
&= \ln S_t + r(T - t) - \frac{1}{2}\rho_H\sigma^2(T - t)^{2H}N_2^{1-2H}, \\
v_{T,t} &= E\left[\left(\ln\left(S_T\right) - m_{T,t}\right)^2|\mathcal{F}_t\right] \\
&= E\left[E\left[\ldots E\left[E\left[\left(\ln\left(S_T\right) - m_{T,t}\right)^2|\mathcal{F}_{t_{N_2-1}}\right]|\mathcal{F}_{t_{N_2-2}}\right]|\ldots\right]|\mathcal{F}_t\right] \\
&= \sum_{j=1}^{N_2} \rho_H\sigma^2\left(t_j - t_{j-1}\right)^{2H} \\
&= \rho_H\sigma^2(T - t)^{2H}N_2^{1-2H}.
\end{aligned}
$$

The distribution of S_T conditioned on the information available at time t is then lognormal with the following moments:

$$
\begin{aligned}
M_{T,t} &= E\left[S_T|\mathcal{F}_t\right] \\
&= \exp\left(m_{T,t} + \frac{1}{2}v_{T,t}\right) \\
&= S_t e^{r(T-t)}, \\
V_{T,t} &= E\left[\left(S_T - M_{T,t}\right)^2|\mathcal{F}_t\right] \\
&= \exp\left(2m_{T,t} + 2v_{T,t}\right) - \exp\left(2m_{T,t} + v_{T,t}\right) \\
&= S_t^2 e^{2r(T-t)}\left[e^{\rho_H\sigma^2(T-t)^{2H}N_2^{1-2H}} - 1\right].
\end{aligned}
$$

Let us take a look at what happens, if N_2, i.e. the fineness of the tradable grid, increases. Recall that for any finite time interval between two tradable points in time—no matter how small—one can exclude arbitrage on the tradable grid by refining it, adding a number of non-tradable in-between steps. However, despite the absence of arbitrage, for $N_2 \to \infty$ and $H > \frac{1}{2}$ the uncertainty of the model disappears:

$$
\lim_{N_2\to\infty} V_{T,t} = \lim_{N_2\to\infty}\left(S_t^2 e^{2r(T-t)}\left[e^{\rho_H\sigma^2(T-t)^{2H}N_2^{1-2H}} - 1\right]\right) = 0.
$$

Consequently, the terminal value of the underlying approaches a deterministic value: It is current value of the stock, compounded with the riskless interest rate. To put it differently, the risky asset coincides with the riskless one.

In search of an explanation, we recall the variance properties of fractional Brownian motion for short time intervals of length τ. We showed that the variance equals τ^{2H}. Consequently, for the persistent parameter domain, we have less randomness in the short run than for the classical case. In the multi-time equilibrium approach, we now imposed restrictions on the behavior of the process applying equilibrium conditions on intervals of decreasing length. Depending on the randomness of the process, this evokes different results. Roughly speaking: While geometric Brownian motion can be forced to become a martingale without losing its randomness, the same restrictions are too strong for the persistent price process and all randomness is eliminated.

6.4 Deeper Insights Provided by Discretization: The Continuous Time Case Reconsidered

In Chap. 3 we considered discrete versions of the fractional Black–Scholes market setting. The relation between the discrete two-time equilibrium approach and the continuous two-time model of Chap. 3 is obvious: With an increasing number of in-between steps, the discrete pricing formula approximates the continuous one. As both models focus on a two-time equilibrium, the convergence is simply due to the respective convergence of the basic processes, i.e., the fact that the market model converges.

Note that the main incentive of choosing a two-time access to the continuous time model was to provide an easy approach ensuring absence of arbitrage. The two-time focus had the following advantage: It allowed us to concentrate on the characteristics of fractional Brownian motion and the impact of persistence on option pricing. However, it was not obvious at first glance why the respective calculations should not be possible for other specifications of the equilibrium conditions.

Recall that also for the discrete multi-time approaches we observed a sort of convergence when refining the scale of the time axis. Driven by this insight, we now turn the tables: We pose the proximate question whether there are also continuous time analogues to these cases and how they look like. In other words, we investigate multi-time equilibrium approaches from a continuous time vantage point.

Concerning the discrete multi-time approaches, we differentiated between the multi-time equilibrium related to current time t and the local multi-time equilibrium. Both concepts led to uniqueness of the so-called equilibrium drift rates. The first approach again ensured convergence to the continuous time pricing formula which we called the fractional Black–Scholes pricing formula. Contrarily, the local equilibrium approach eliminated uncertainty and yielded the 'deterministic' pricing formula which we already obtained when discussing aspects of predictability and arbitrage in Chap. 4.

In the first concept, equilibrium conditions were postulated to hold with respect to current time t. As a reminder, for all discrete $t < s < T$, we introduced the equilibrium conditions

$$E\left[S_s | \mathcal{F}_t\right] = S_s \left(1 + r/n\right)^{(s-t)n}. \tag{6.11}$$

This implied a deterministic specification of the equilibrium drift taking piece-wise constant values between two tradable nodes. The approach turned out to be an extension as well as a refinement of the two-time equilibrium making possible, for example, the valuation of American options. Like in the two-time approach, the European option formula tends to the fractional Black–Scholes formula if the tradable grid gets finer.

If we transfer this idea to the continuous setting, we get an infinite number of equilibrium conditions of the form

$$E\left[e^{-r(s-t)} S_s | \mathfrak{F}_t^H\right] = S_t, \qquad \forall s \in [t, T]. \tag{6.12}$$

Note that, although consecutive transactions are imposed to have a minimal time interval in-between, it is still necessary to postulate the equilibrium for the infinite amount of points in time. This is due to the fact that ex ante no point in time can be excluded from trade. Restrictions not occur until an investor proceeds his first transaction. Then, of course, he is faced with a certain waiting period or delay. In our system of equilibrium conditions, current time t is fixed, whereas s takes all values between t and maturity T. Hence, this system of conditions does not turn the stock process into a martingale: The condition will not hold for an arbitrary combination of points in time s_1, s_2 with $t < s_1, s_2 < T$.

We now apply the same idea as in the discrete consideration before. In order to satisfy the whole system of equilibrium conditions, we allow for a time-varying, but deterministic equilibrium drift. In this case, the fractional Itô theorem still holds, and the same calculations as in Chap. 5 lead to the following system of conditions for the equilibrium drift rate $\bar{\mu}$:

$$\int_t^s \bar{\mu}_u \, du = r(s - t) - \sigma \hat{\mu}_{s,t}, \qquad \forall s \in [t, T], \tag{6.13}$$

where $\hat{\mu}_{s,t}$ is the conditional mean of B_s^H based on information in time t. This equation implicitly determines the process of the equilibrium drift $\bar{\mu}_u$. Meanwhile, like the term $\bar{\mu}(T - t)$ in the continuous two-time approach, the term $\int_t^s \bar{\mu}_u \, du$ vanishes, as soon as we insert it into the conditional moments of the stock price process. Consequently, the formulae (5.26) and (5.27) of the conditional mean m and the conditional variance v do not change and the option pricing formula also remains unchanged.

Finally, we consider the continuous time analogy to the local multi-time equilibrium concept. In our discrete time framework, the idea was to introduce equilibrium conditions for each pair of consecutive tradable nodes. The transition to the continuous time point of view results in the following system of equilibrium conditions:

$$E\left[e^{-r\delta}S_{s+\delta}|\mathfrak{F}_s^H(\omega)\right] = S_s, \qquad \forall s \in [t,T), \omega \in \Omega. \tag{6.14}$$

The parameter δ denotes the—once chosen—minimal amount of time between two consecutive transactions which we need to exclude arbitrage (see Sect. 4.4). Like in the case looked at before, we need to fulfill this condition for all points in time. More than this, the conditions also need to hold for all states of nature which are possible at the respective points in time. The equilibrium drift for the next time step then depends on the historical path. Hence, like in the discrete local equilibrium setting, we have to allow for a stochastic or path-dependent specification of the equilibrium drift rate.

If one wants to apply the respective steps of our continuous time model and tries to exploit the system of equilibrium conditions, one becomes aware of the fact that the fractional Itô theorem is no longer applicable: The theorem does not hold if we state a time-dependent and state-dependent drift rate taking values conditional on the realized path. In particular, we are not able to derive a formula comparable to (6.13) which would enable us to eliminate the drift like in the other two cases we discussed before. As a straightforward consequence, also the derivation of the fractional Black–Scholes type option pricing formula no longer holds. Contrariwise, the discrete setting clarified that the state-dependent mean correction on decreasing time scales affects the distribution of the process and leads to a process without any uncertainty where option values are nothing but their discounted inner value.

We summarize the considerations of this chapter: Concerning the calculations of the option pricing problem in Sect. 5.4, there is only one crucial step that is needed to exploit the system of equilibrium conditions, which is the fractional Itô theorem. The different specifications of this equilibrating system (two-time, multi-time with respect to t, local multi-time) necessitate different specifications of the equilibrium drift (constant, deterministic, path-dependent). The first case is our basic model and applying the fractional Itô theorem for a constant drift is trouble-free. Also the deterministic case did not cause any problems: The theorem and as a consequence also the pricing formula still hold. For the third case, however, the theorem and the further calculations of our basic model cannot be applied. Moreover, the parallel to the respective discrete time model clarifies: The equilibration of the process on the very small time scale by a path-dependent drift instantaneously counterbalances the stochastic of the process and thereby eliminates it.

Chapter 7
Conclusion

This thesis dealt with the suitability of the stochastic process of fractional Brownian motion when modeling randomness in financial market settings.

In the preliminary Chap. 2 we pointed out why fractional Brownian motion could be an interesting candidate for financial models. We showed that it combines the possibility to capture serial correlation of a stochastic process with a good analytical treatability for still being Gaussian. As only one additional parameter was introduced, fractional Brownian motion turned out to be a parsimonious extension of classical Brownian motion. The range of this so-called Hurst parameter H could be divided into three cases. For $0 < H < \frac{1}{2}$, we observed anti-persistent behavior, for $\frac{1}{2} < H < 1$ persistence occurred. When H equalled $\frac{1}{2}$, we obtained the classical Brownian case. This property allowed for an easy benchmark concerning all the results that should be derived throughout the rest of the thesis: The new, generalized result of the fractional Brownian world should always include the corresponding well-known result of classical Brownian motion.
Further parallels between fractional and classical settings were stressed concerning integration calculus. Both the Stratonovich and the Itô calculus of Brownian integration theory, found its equivalents in the fractional context. We highlighted the basic idea as well as the main results of fractional integration calculus including some technicalities. Most importantly, we recalled a fractional version of the classical Itô theorem which had been recently provided by Duncan et al. (2000).

In Chap. 3, we considered a discretization of fractional Brownian motion by a binomial process. Based on the work of Sottinen (2001) who had provided a discrete approximation of arithmetic fractional Brownian motion, we visualized the procedure by depicting binomial trees. Moreover, we investigated the conditional properties of the binomial fractional Brownian motion and thereby emphasized the influence of the historic path on the future distribution. We proceeded and looked at a fractional binomial price process represented by a binomial version of a geometric fractional Brownian motion. By visualization of these processes, we could get a first impression of the key problem being immanent in financial models based on fractional Brownian motion. Due to its path-dependence, some paths of the binomial tree led to situations where at a certain node (which is a state of nature at a certain point in time) both subsequent nodes yielded a payoff that exceeded the return of the riskless asset. Applying a simple one-step buy-and-hold strategy, this again offered the possibility of a riskless gain without initial capital, i.e. an arbitrage. The solution to the problem that excluded arbitrage, was the following: Market participants were assumed to be subject to restrictions concerning the speed of adapting their trading strategy, or more simply: A single investor cannot be as fast the market. Hence, we fixed a small time interval that investors cannot go below, but continued to refine the discretization of the process. The in-between evolution of the stock then provided a sufficient degree of volatility and made arbitrage strategies impossible. This ensured the reasonability of discrete time financial models based on a binomial fractional Brownian motion.

In the following chapter, we readdressed ourselves to the continuous time process of fractional Brownian motion. Stimulated by the insights of the binomial setting, we investigated in Chap. 4 a financial market setting consisting of a riskless asset as well as a risky one driven by a geometric fractional Brownian motion. We first recalled the debate of the history: The first theoretical results by Delbaen and Schachermayer (1994) as well as by Rogers (1997) had suggested a categorical rejection of fractional market models for reasons of arbitrage. Though the introduction of Wick–Itô calculus had inspired some promising results (Hu and Øksendal (2003) and Elliott and van der Hoek (2003)), Bjork and Hult (2005) had shown that their implied assumptions were economically meaningless.

We then worked out that as long as the possibility of continuous tradability exists, the predictability of fractional Brownian motion always eliminates randomness from option pricing. We stated a fractional inversion of the work of Sethi and Lehoczky (1981): They had shown for the classical case that—if applied sensibly—both Stratonovich and Itô calculus lead to the Black–Scholes pricing formula for a European call option. Contrariwise, we showed that the correct usage and interpretation of both pathwise and Wick–Itô calculus does not lead to an option pricing formula à la Black–Scholes but to a formula where the price is nothing but the maximum of the discounted inner value and zero. On the one hand, our considerations provided another example concerning the importance of the correct interpretation of integration concepts as disregarding them leads to results like those by Hu and Øksendal (2003). On the other hand, we drew the following conclusion: In order to ensure a reasonable pricing and absence of arbitrage in the fractional Brownian market model, we had to restrict trading strategies to be non-continuous. This was the logical counterpart to the phenomenon we had observed in the discrete time setting. Furthermore, it was perfectly in line with similar findings received by Cheridito (2003).

Albeit the restriction to non-continuous trading strategies ensured absence of arbitrage, the non-continuity still ruled out the common arbitrage pricing approach based on dynamic hedging. In Chap. 5 we suggested a preference based pricing approach which allowed us to renounce continuous tradability and to focus on two points in time, which were the present and maturity. An equilibrium condition relating these two points in time was exploited. It postulated that—as we asserted well-informed and risk-neutral investors— the conditional expected payoff of the underlying should equal the respective certainty equivalent.

This approach made it necessary to evaluate the historical information from the path of the stock price process, i.e. we introduced the conditional distribution of fractional Brownian motion. By means of this, we derived that there can only be one unique constant drift rate of the underlying for which the desired equilibrium is given. This drift rate was composed of two parts: the riskless interest rate plus a correction accounting for the evolution of the historical path.

We went on to price European options by their conditional expected payoff. The derived formulae draw their attractiveness from the fact that the fractional pricing model includes the traditional Markovian case of classical Brownian motion. So, the existing parallels enhanced the understanding of fractional option pricing. Moreover the analysis of the partial derivative with respect to the Hurst parameter made it possible to point out the fractional particularities of the formulae. By name, these were the variance-based narrowing and power effects which accorded with the economic intuition concerning the idea of persistence and antipersistence. The analysis of the term structure of implied volatilities showed that our model yielded non-flat curves that could be derived analytically. Therefore, it is in principle suited to explain real market phenomena like volatility smirks over time to maturity.

In our next step we translated the continuous time pricing model into the discrete time binomial setting that we had introduced before. We started Chap. 6 by focusing on a two-time equilibrium. For the latter, we presented the idea of relative pricing, where equilibrium is realized by a change of measure as well as that of absolute pricing, where equilibrium is achieved by an drift adjustment of the underlying. Both led to the same option prices. With an increasing level of discretization, the option prices again tended to the value given by the continuous time pricing formula.

Motivated amongst other things by the desire to price also American options, we expanded the points of equilibrium onto the whole tradable grid. Concerning these multi-time equilibrium approaches, we discussed only the idea of absolute pricing. We differentiated between two of these multi-time approaches: In the first approach, equilibrium was postulated to hold for each tradable node when relating it to current time t. The second one which we called local multi-time equilibrium, introduced equilibrium relations for each multiple consisting of one node and its immediate successors on the tradable grid. While the first concept also yielded convergence to the continuous time pricing formula, we proved that the postulation of a local multi-time equilibrium eliminated the uncertainty of the underlying process. The option value then tended to the maximum of the discounted inner value and zero.

Finally, in Sect. 6.4, we factored the new insights into a reconsideration of the continuous time model and brought up the question of whether equivalents of the multi-time approaches could also be constructed in the continuous framework. We emphasized the strong relation between the design of the equilibrium condition, on the one hand, and the shape of the resulting equilibrium drift, on the other hand. The latter however turned out to play a crucial role if one wants to adopt the discrete time procedure in the continuous setting. The local multi-time equilibrium implied a stochastic equilibrium drift rate which made the application of the fractional Itô theorem impossible. The other two approaches led to a constant or deterministic drift, respectively, so the theorem could be applied.

On the whole, the nature of fractional Brownian motion brings about a reduction of short-run uncertainty when comparing it with the classical Brownian approach. Although predictability can lead to arbitrage possibilities or eliminate randomness, the problem can be solved in an incomplete framework by applying risk preference based pricing approaches based on suitable equilibrium arguments. Based on these important conceptual results, the main achievement of this thesis is a most tangible one: The closed form pricing formulae for European options in the continuous time fractional Brownian market.

The results in this thesis were derived under the assumption of risk-neutral investors. Further research could deal with different assumptions concerning risk preferences. The basic idea should be to introduce a different kind of equilibrium condition accounting for the respective certainty equivalent.
Furthermore, one could discuss financial models where the underlying follows combined processes that include fractional Brownian motion as one building block such as a fractional Lévy motion (see e.g. Huillet (1999)).

We conclude with the statement that fractional Brownian motion is by no means an absurd candidate for financial models. As soon as one stops clinging to dynamical completeness, fractional Brownian motion offers convenient properties: One can parsimoniously introduce serial correlation into financial models and nevertheless get closed-form solutions that are easy to handle and in line with economic intuition.

References

Abramowitz, M.A., Stegun, I.A. (1972): Handbook of Mathematical Functions. New York: Dover Publications.

Ash, R.B. (1972): Real Analysis and Probability. Academic Press, New York.

Barth, W. (1996): Fraktale, Long Memory und Aktienkurse – eine statistische Analyse für den deutschen Aktienmarkt. EuL-Verlag, Bergisch-Gladbach, Köln.

Bender, C. (2003a): Integration with respect to Fractional Brownian Motion and Related Market Models. University of Konstanz, Department of Mathematics and Statistics: Ph. D. thesis.

Bender, C. (2003b): An S-Transform approach to integration with respect to a fractional Brownian motion, Bernoulli 9(6), p. 955–983.

Bender, C. (2003c): An Itô Formula for Generalized Functionals of a Fractional Brownian Motion with arbitrary Hurst parameter, Stoch Proc Appl 104, p. 81–106.

Bender, C., Elliott, R.J. (2004): Arbitrage in a Discrete Version of the Wick-Fractional Black-Scholes Market. Math Oper Res 29, p. 935–945.

Bender, C., Sottinen, T., Valkeila, E. (2006): Arbitrage with fractional Brownian motion?, Theory of Stochastic Processes 12(28).

Benth, F.E. (2003): On arbitrage-free pricing of weather derivatives based on fractional Brownian motion, Appl Math Financ 10(4), p. 303–324.

Billingsley, P. (1968): Convergence of Probability Measures. Wiley, New York.

Bjork, T., Hult, H. (2005): A note on Wick products and the fractional Black-Scholes model, Financ Stochast 9(2), p. 197–209.

Black, F., Scholes, M. (1973): The Pricing of Options and Corporate Liabilities, J Polit Econ 81, p. 637–654.

Brennan, M.J. (1979): The Pricing of Contingent Claims in Discrete Time Models, J Financ 24(1), p. 53–68.

Carr, P., Wu, L. (2003): The Finite Log Stable Process and Option Pricing, J Financ 58(2), p. 753–777.

Cheridito, P. (2001a): Regularizing fractional Brownian motion with a view towards stock price modelling. Eidgenöss Technische Hochschule Zürich, Swiss Federal Institute of Technology: Ph. D. Thesis.

Cheridito, P. (2001b): Mixed fractional Brownian motion, Bernoulli **7**, p. 913–934.

Cheridito, P. (2003): Arbitrage in fractional Brownian motion models, Finance Stochast **7**(4), p. 533–553.

Cont, R., Tankov, P. (2004): Financial Modelling with Jump Processes. Chapman and Hall, CRC Press.

Cox, J., Ross, S., Rubinstein, M. (1979): Options pricing: A simplified approach, J Financ Econ **7**, p. 229–263.

Cox, J., Rubinstein, M. (1985): Options markets. Prentice Hall, New York.

Dasgupta, A. (1998): Fractional Brownian motion: its properties and applications to stochastic integration. University of North carolina, Ph. D. Thesis.

Dasgupta, A., Kallianpur, G. (2000): Arbitrage opportunities for a class of Gladyshev processes, Appl Math Opt **41**, p. 377–385.

Delbaen, F., Schachermayer, W. (1994): A General Version of the Fundamental Theorem of Asset Pricing, Math Ann **300**, p. 463–520.

Derman, E., Kani, I. (1994): Riding on a smile, Risk, **7**(2), p. 32–39.

Duncan, T.E., Hu, Y., Pasik-Duncan, B. (2000): Stochastic Calculus for Fractional Brownian Motion, SIAM J Control Optim **38**(2), p. 582–612.

Dupire, B. (1994): Pricing with a smile, Risk, **7**(1), p. 18–20.

Elliott, R.J., Van Der Hoek, J. (2003): A General Fractional White Noise Theory and Applications to Finance, Math Financ **13**(2), p. 301–330.

Fama, E.F. (1965): The behavior of stock market prices, J Bus **38**, p. 34–105.

Genest, C., Ghoudi, K., Rémillard, B. (1996): A note on tightness, Stat Probabil Lett **27**(4), p. 331–339.

Gripenberg, G., Norros, I. (1996): On the prediction of Fractional Brownian Motion, J Appl Probab **33**, p. 400–410.

Guasoni, P. (2006): No arbitrage under transaction cost, with fractional Brownian motion and beyond, Math Financ **16**(3), p. 569–582.

Holden, H., Øksendal, B., Ubøe, J., Zhang, T. (1996): Stochastic Partial Differential Equations. A modeling, white Noise Functional Approach. Birkhäuser, Boston.

Hu, Y., Øksendal, B. (2003): Fractional White Noise Calculus and Applications to Finance, Infin Dimens Anal Qu **6**(1), p. 1–32.

Huillet, T. (1999): Fractional Lévy motions and related processes, J Phys A - Math Gen, **32**, p. 7225–7248.

Itô, K. (1951): Multiple Wiener Integrals, J Math Soc Jpn **3**, p. 157–169.

Lin, S.J. (1995): Stochastic Analysis of Fractional Brownian Motion, Stoch Stoch Rep **55**, p. 121–140.

Lo, A.W., MacKinley, A.C. (1988): Stock market prices do not follow random walks: Evidence from a simple specification test, Rev Financ Stud **1**, p. 41–66.

Mandelbrot, B.B., van Ness, J.W. (1968): Fractional Brownian Motions, Fractional Noises and Applications, SIAM Rev **10**(4), p. 422–437.

Merton, R. (1973): Theory of Rational Option Pricing, Bell J Econ Manage Sci **4**, p. 141–183.

Musiela, M., Rutkowskii, M. (2005): Martingale Methods in financial Modelling. Springer, Berlin-Heidelberg-New York.

Necula, C. (2002): Option Pricing in a Fractional Brownian Motion Environment. Preprint, Academy of Economic Studies, Bucharest.

Nelson, D., Ramaswamy, K. (1990): Simple binomial processes as diffusion approximations in financial models, Rev Financ Stud **3**(3), p. 393–430.

Norros, I., Valkeila, E., Virtamo, J. (1999): An elementary approach to a Girsanov formula and other analytical results on fractional Brownian motion, Bernoulli **5**, p. 571–587.

Nuzman, C.J., Poor, H.V. (2000): Linear Estimation of Self-Similar Processes via Lamperti's Transformation, J Appl Probab **37**, p. 429–452.

Øksendal, B. (1996): An Introduction to Malliavin Calculus with Applications to Economics, Lecture Notes, Norwegian School of Economics and Business Administration.

Øksendal, B. (2006): Fractional Brownian Motion in Finance, in: Jensen(ed.): Stochastic Economic Dynamics. Cambridge University Press.

Rogers, L.C.G. (1997): Arbitrage with fractional Brownian motion, Math Financ **7**(1), p. 95–105.

Rostek, S., Schöbel, R. (2006): Risk preference based option pricing in a fractional Brownian market, Tübinger Diskussionsbeitrag 299.

Sethi, S.P., Lehoczky, J.P. (1981): A Comparison of the Itô and Stratonovich Formulations of Problems in Finance, J Econ Dyn Control **3**, p. 343–356.

Shiryayev, A.N. (1998): On arbitrage and replication for fractal models, research report **20**, MaPhySto, Department of Mathematical Sciences, University of Aarhus, Denmark.

Sottinen, T. (2001): Fractional Brownian Motion, Random Walks and Binary Market Models, Financ Stochast **5**(3), p. 343–355.

Sottinen, T., Valkeila, E. (2003): On arbitrage and replication in the fractional Black-Scholes pricing model, Stat Dec **21**, p. 93–107.

Stratonovich, R.L. (1966): A new representation for stochastic integrals and equations, SIAM J Control **4**, p. 362–371.

Willinger, W., Taqqu, M.S., Teverovsky, V. (1999): Stock market prices and long-range dependence, Financ Stochast **3**, p. 1–13.

Zähle, M. (1998): Integration with respect to Fractal Functions and Stochastic Calculus I, Probab Theory Rel **111**, p. 333–374.